30-SECOND
DATA SCIENCE

30-SECOND
DATA SCIENCE

The 50 key principles and innovations
in the field of data-gathering, each
explained in half a minute

Editor
Liberty Vittert

Contributors
Maryam Ahmed
Vinny Davies
Sivan Gamliel
Rafael Irizarry
Robert Mastrodomenico
Stephanie McClellan
Regina Nuzzo
Rupa R. Patel
Aditya Ranganathan
Willy Shih
Stephen Stigler
Scott Tranter
Liberty Vittert
Katrina Westerhof

Illustrator
Steve Rawlings

IVY PRESS

First published in the UK in 2020 by
Ivy Press
An imprint of The Quarto Group
The Old Brewery, 6 Blundell Street
London N7 9BH, United Kingdom
T (0)20 7700 6700
www.QuartoKnows.com

British Library Cataloguing-in-
Publication Data
A catalogue record for this
book is available from the
British Library.

ISBN: 978-1-78240-892-5

This book was conceived,
designed and produced by
Ivy Press
58 West Street, Brighton BN1 2RA, UK

Publisher **David Breuer**
Editorial Director **Tom Kitch**
Art Director **James Lawrence**
Commissioning Editor **Natalia Price-Cabrera**
Project Editor **Caroline Earle**
Designer **Ginny Zeal**
Illustrator **Steve Rawlings**
Glossaries **Maryam Ahmed**

Printed in China

10 9 8 7 6 5 4 3 2 1

CONTENTS

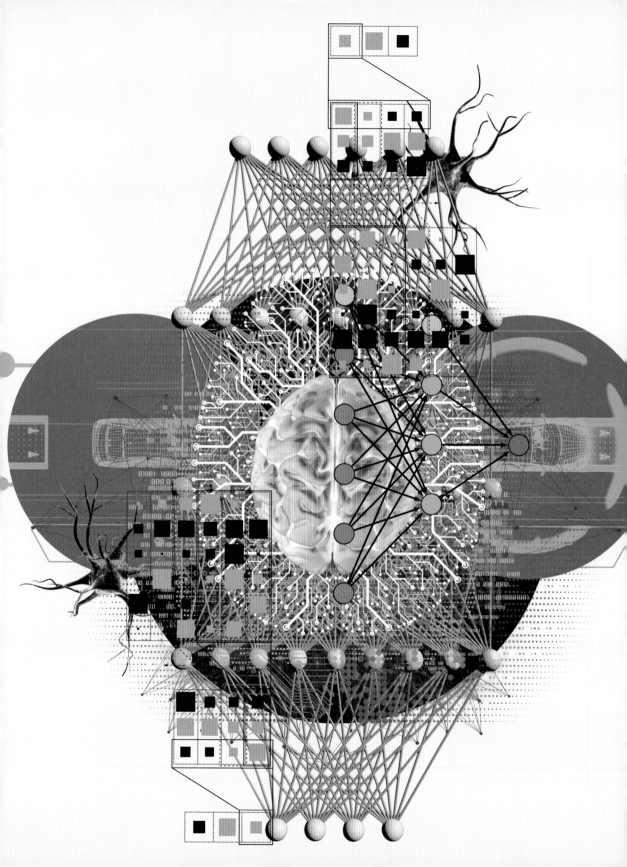

FOREWORD
Xiao-Li Meng

"If you want to solve all the problems in the world,
major in computer science." When a speaker at an AI conference displayed
this line, my statistical ego got provoked immediately to almost six-sigma
level. Thankfully, the next line appeared in less than three seconds: "If you
want to solve all the problems created by computer science, enrol in a
graduate program in the Faculty of Arts of Sciences."

Whoever humoured us with this clever pairing has a profound
understanding of the brilliant and bewildering time we live in. Advances
in computer science and technology created the digital age, which in turn
created data science. Nothing seems beyond reach when we can have so
much data to reveal the secrets of nature and its most advanced species.

But there is no free lunch – a universal law of data science (and of life).
Here are just a couple of paradise-paradox pairings of food for thoughts.
Personalized medicine surely sounds heavenly. But where on earth can they
find enough guinea pigs for me? Undoubtedly, we need to collect as much
data as possible about humans in order to advance the AI technologies.
But please only study other people – don't you dare invade my privacy!

For those who can still enroll in a graduate program and afford at
least 31,536,000 seconds, do it as if there is no next 30 seconds. For
those who cannot, this book takes 50 x 30 seconds, give or take your
personal six sigma. Completing this book will not make you a 30-second
data scientist. But without understanding its content, your application
for the digital-age citizenship will be denied with 99% certainty. You can
take a chance, of course.

INTRODUCTION
Liberty Vittert

We have lived as humanists for a long time, using our instinct, with our thoughts, opinions and experience driving our decisions. However, now we are moving into the era of Dataism – letting data drive our every decision. From climate change to the refugee crisis to healthcare, data is a driving force, and not just in these all-encompassing issues but also in our daily lives. Instead of going to a bookshop, Amazon can tell you what you want to read. Likewise, dating apps will tell you who you are compatible with, using endless amounts of collected data.

Humanism and Dataism are currently pushing back against each other. Some people want to be purely data-driven, others don't want to let go of the human touch. Data science, as a discipline, brings both humanism and Dataism together. It combines the vast databases, powerful statistical tools that drive computing processes, and analysis together with the common sense and quantitative reasoning we as humans have been developing over thousands of years. Data Science does not mean just being data-driven or human-driven: it is the art of both together.

Before we begin detailing this book, let's take a step back in time to the seventeenth century and visit Blaise Pascal, a French monk with a crisis of faith. He decided to think about his future options with the information he had available to him – the data, if you will:

If God doesn't exist and I believe in Him then I might have a wasted life with false belief, but nothing happens.

If God doesn't exist and I don't believe in Him then I didn't waste my life with false belief, but again, nothing happens.

If God does exist and I do believe in Him then I have a wonderful eternity in Heaven.

But if God does exist and I don't believe in Him then it's eternal hell-fire for me.

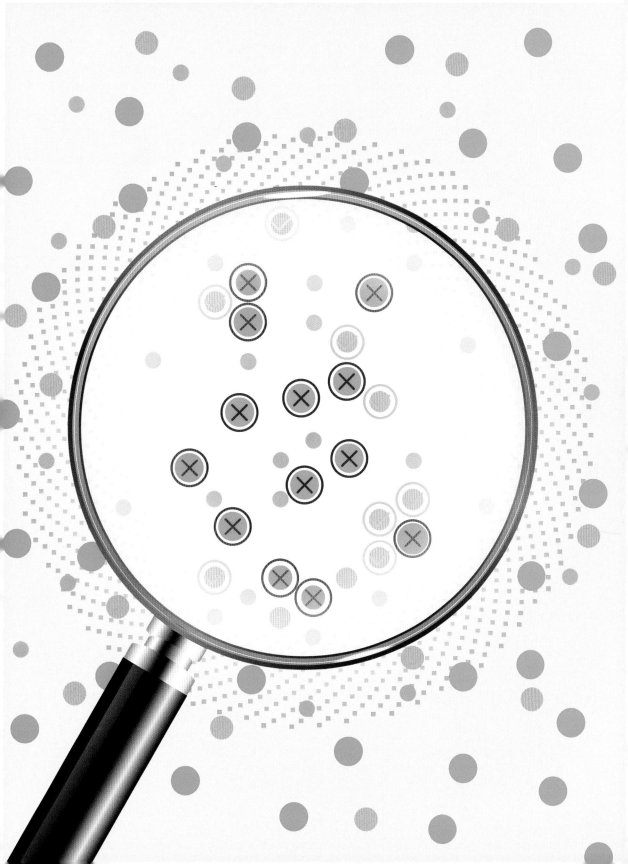

Pascal used the data he had to make a decision to optimize his future happiness and mitigate potential risk. Really, that is what data science is: taking past and current information in order to predict the likelihood of future events, or, rather, the closest thing to a crystal ball that the world has at its disposal. The only difference between us and Pascal is that we live in a world with far more than four bits of data to analyse; we have endless amounts.

It is estimated that we produce over 2.5 exabytes of data per day. A quick calculation makes that the same amount of information as stacking Harry Potter books from the Earth to the Moon, stacking them back, and then going around the circumference of the Earth 550 times. And that is simply the amount of data produced per day!

How the book works
The first two chapters break down data science into its basic elements, followed closely by its most important and also most under-discussed facet – what it cannot tell us. The following five chapters venture into how data science affects us in every way – science, society, business, pleasure and our world's future. Within each individual topic there is: a **3-Second Sample**, an insightful glimpse into the topic; followed by the more detailed **30-Second Data** explanation; and finally the **3-Minute Analysis**, which provides the reader with an opportunity to take a deeper dive into particular complexities and nuances of the discussion at hand.

This book has been carefully compiled by industry-specific experts to help guide us through how data is changing every industry and every facet of our lives in ways we haven't even imagined, while clearly showing the quantitative reasoning and ethical quandaries that come with the dawn of any new era.

BASICS ◑

algorithm Set of instructions or calculations designed for a computer to follow. Writing algorithms is called 'coding' or 'computer programming'. The result of an algorithm could be anything from the sum of two numbers to the movement of a self-driving car.

automation Repetitive tasks or calculations carried out faster and more efficiently by computers than by humans.

Bayesian analysis Statistical method for estimating probabilities based on observed data and prior knowledge, making it possible to answer questions such as 'What is the probability that a person will develop lung cancer if they are a smoker?'

binary Representation of information as a string of 1s and 0s – an easy format for computers to understand, and so is fundamental to modern computing.

bivariate analysis Restricted to one output or dependent variable.

causal inference Determining whether a change in one variable directly causes a change in another variable. For example, if increased coffee consumption was found to directly cause an improvement in exam scores, the two variables would have a causal relationship.

cluster A group of computers working in parallel to perform a task. It is often more efficient for complex computation tasks to be performed on a cluster rather than a single computer.

core (computer/machine) Central processing unit (CPU) of a computer, where instructions and calculations are executed; communicates with other parts of a computer. Many modern computers use multi-core processors, where a single chip contains multiple CPUs for improved performance.

data set A set of information stored in a structured and standardized format; might contain numbers, text, images or videos.

Enigma code Method of scrambling or encrypting messages employed by the German armed forces during the Second World War which was cracked by Alan Turing and his colleagues at Bletchley Park.

epidemiology The study of the incidence of health conditions and diseases, which populations are most vulnerable and how the associated risks can be managed.

interpretability Extent to which a human can understand and explain the predictions or decisions made by a mathematical model.

model/modelling Real world processes or problems in mathematical terms; can be simple or very complex, and are often used to make predictions or forecasts.

multivariate analysis Measures the effect of one or more inputs, or independent variables, on more than one output, or dependent variable. For example, a study that models the effect of coffee consumption on heart rate and blood pressure would be a multivariate analysis.

normal (Gaussian) distribution Bell-shaped curve describing the spread or distribution of data across different values. Data sets that are often normally distributed include exam scores, the heights of humans and blood pressure measurements. Normal distribution shows the probability of a random variable taking different values. Many statistical analyses assume the data is normally distributed.

statistical association The strength of relationship between two variables or measurements, e.g. there is an association between human age and height. One commonly used measure of association is Pearson's correlation coefficient.

terabyte A measure of a computer or hard drive's storage capacity, abbreviated to TB. One TB is equal to one trillion bytes.

DATA COLLECTION

the 30-second data

Data science was born as a subject when modern computing advances allowed us to suddenly capture information in huge amounts. Previously, collecting and analysing data was limited to what could done by hand. Modern advances now mean that information is collected in every part of our lives, from buying groceries to smart watches that track every movement. The vast amount now collected is set to revolutionize every aspect of our lives, and massive companies have emerged that collect data in almost unimaginable quantities. Facebook and Google, to name just a couple, collect so much information about each of us that they could probably work out things about us that even closest friends and family don't know. Every time we click on a link on Google or like a post on Facebook, this data is collected and these companies gain extra knowledge about us. Combining this knowledge with what they know about other people with similar profiles to ourselves means that these companies can target us with advertising and predict things about us that we would never think possible, such as our political allegiances.

3-SECOND SAMPLE
Since the invention of modern computing, 'big data' has become a new currency, helping companies grow from conception to corporate giants within a decade.

3-MINUTE ANALYSIS
The amount of data that we now collect is so massive that the data itself has its own term – big data. The big data collected in the modern era is so huge that companies and researchers are in a constant race to keep up with the ever-increasing requirements of data storage, analysis and privacy. Facebook supposedly collects 500+ terabytes of data every day – it would take over 15,000 MacBook Pros per day to store it all.

RELATED TOPICS
See also
TOOLS
page 22

SURVEILLANCE
page 82

REGULATION
page 150

3-SECOND BIOGRAPHIES
GOTTFRIED LEIBNIZ
1646–1716
Helped develop the binary number system, the foundation of modern computing.

MARK ZUCKERBERG
1984–
Co-founded Facebook with his college room-mates in 2004, and is now CEO and chairman.

30-SECOND TEXT
Vinny Davies

Personal data has become the sought-after commodity of the technical age.

HOW WE VISUALIZE DATA

the 30-second data

Where does the quote 'ninety per cent of politicians lie' come from, and more importantly, is it true? In everyday life, summaries of data can be seen in many forms, from pie charts telling us the nation's favourite chocolate bar, to news articles telling us the chance of getting cancer in our lifetime. All these summaries come from, or are based on, information that has been collected, but so often summaries seem to contradict each other. Why is this the case? Well, data isn't usually simple and nor is summarizing it; I may summarize it one way, you another. But who is right? Therein lies the problem: it is possible to be manipulated by the data summaries we are shown. Even summaries that are true may not provide information that is a fair and accurate representation of the data which that summary represents. For instance, did you know that teenage pregnancies dramatically reduce when girls reach 20 years of age? Technically true, but realistically not a useful summary. The next time you see a summary, think about how it could have been manipulated, and then consider the results of the summary accordingly.

RELATED TOPICS
See also
LEARNING FROM DATA
page 20

CORRELATION
page 42

VOTE SCIENCE
page 90

3-SECOND SAMPLE
Data is everywhere in everyday life, but most of us don't work in data science; so how is that data seen and what beliefs are formed from it?

3-MINUTE ANALYSIS
Physically visualizing the massive amounts of complex data collected is a challenge in itself. Most modern data sets are almost impossible to visualize in any sensible way and therefore any visual summaries are usually a very simplified interpretation of the data. This also means that visual summaries can easily be misrepresented, and what is seen isn't always as straightforward as it seems.

3-SECOND BIOGRAPHIES
BENJAMIN DISRAELI
1804–81
Former British Prime Minister to whom the quote 'there are three types of lies: lies, damned lies and statistics' is often attributed.

STEPHAN SHAKESPEARE
1957–
Co-founder and CEO of opinion polls company YouGov, which collects and summarizes data related to world politics.

30-SECOND TEXT
Vinny Davies

In the realm of data science, seeing is not necessarily believing – it's always wise to look beyond the summary presented.

LEARNING FROM DATA

the 30-second data

3-SECOND SAMPLE
Analysing and modelling data can highlight information not obvious in data summaries, revealing anything from social media trends to causes of cancer.

3-MINUTE ANALYSIS
Learning from data is not a modern phenomenon. In 1854, during an outbreak of cholera in London, Dr John Snow collected and used data to show the source of the disease. He recorded where cholera cases occurred and used the data to map them back to the Broad Street Pump. Residents then avoided the pump, helping to end the outbreak of the disease. The pump remains as a landmark in London to this day.

Collecting data is all very well, but once collected, can more be done with it than just summarizing it? Using models, attempts can be made to gain information from the data in a more complex and useful way than before. Models effectively allow data scientists to use one or more pieces of data to predict an outcome (another piece of data) in which they are interested. For instance, age and gender data could be used to predict whether someone will get arthritis in the next five years. Creating a model with age and gender from previous individuals (knowing whether they got arthritis or not) allows us to predict what could happen to a new individual. As well as simply trying to predict future data, data can also be used to try to establish the cause of a particular outcome. This process is called 'causal inference' and is often used to help understand disease, for example via analysing DNA. However, even though both examples mentioned are trying to predict cases of arthritis, the modelling problems they represent are subtly different and are likely to require vastly different modelling processes. Choosing the best model based on the data and aims associated with a particular project is one of the major skills all data scientists must have.

RELATED TOPICS
See also
DATA COLLECTION
page 16

STATISTICS & MODELLING
page 30

MACHINE LEARNING
page 32

3-SECOND BIOGRAPHIES
JOHN SNOW
1813–58
British physician considered the founding father of epidemiology who is known for tracing the source of a cholera outbreak in London in 1854.

ALAN TURING
1912–54
British mathematician who used data from messages to help crack the Enigma code in the Second World War.

30-SECOND TEXT
Vinny Davies

Once gathered, data can be put through modelling processes, which can enhance understanding.

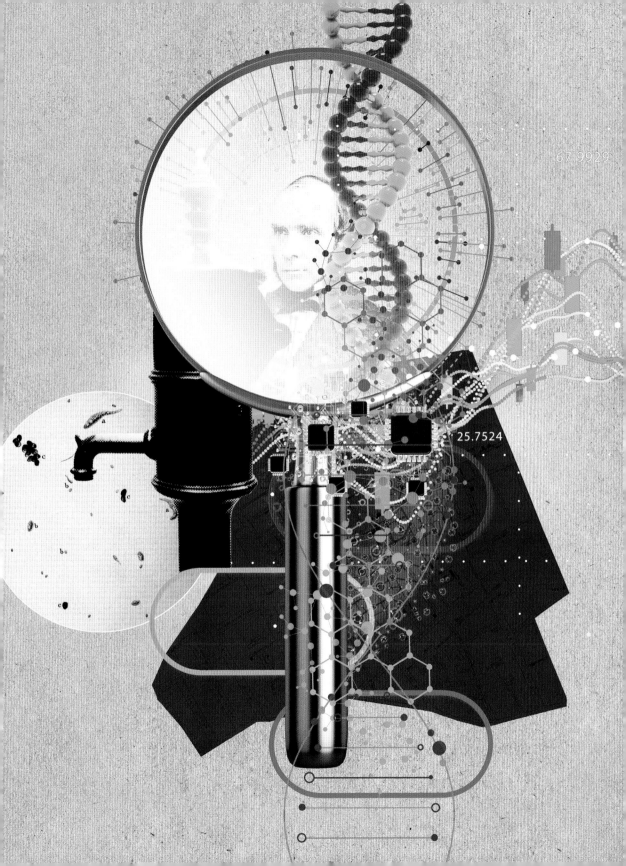

TOOLS

the 30-second data

Dealing with the massive

data sets that are collected and the complex processes needed to understand them requires specialist tools. Data scientists use a wide variety of tools to do this, often using multiple different tools depending on the specific problem. Most of these tools are used on a standard computer, but in the modern era of cloud computing, work is beginning to be done on large clusters of computers available via the internet. A lot of large tech companies offer this service, and these tools are often available to data scientists. In terms of the more standard options in a data scientist's toolbox, they can generally be divided into tools for managing data and tools for analysing data. Often, data is simply stored in spreadsheets, but sometimes, when data gets larger and more complex, better solutions are required, normally SQL or Hadoop. There is a much larger variety of tools used for analysing data, as the methods used often come from different communities, for instance statistics, machine learning and AI, with each community tending to use different programming languages. The most common programming languages used to analyse data tend to be R, Python and MATLAB, although often data scientists will know multiple languages.

RELATED TOPICS
See also
DATA COLLECTION
page 16

LEARNING FROM DATA
page 20

STATISTICS & MODELLING
page 30

3-SECOND SAMPLE
Data is big, models are complex, so data scientists have to use all the computational tools at their disposal. But what are these tools?

3-MINUTE ANALYSIS
While not explicitly a tool in the same sense as Python, SQL, etc, parallel computing is an important part of modern data science. When you buy a computer, you will likely have bought either a dual or quad core machine, meaning that your computer is capable of processing two or four things simultaneously. Many data science processes are designed to use multiple cores in parallel (simultaneously), giving faster performance and increased processing capabilities.

3-SECOND BIOGRAPHIES
WES MCKINNEY
1985–
Python software developer who founded multiple companies associated with the development of Python.

HADLEY WICKHAM
fl. 2006–
Researcher and Chief Scientist at RStudio, known for the development of a number of key tools within the R programming language.

30-SECOND TEXT
Vinny Davies

Data scientists will choose a tool or programming language to suit the task at hand.

REGRESSION

the 30-second data

Regression is a method used to explain the relationship between two or more measurements of interest, for example height and weight. Based on previously collected data, regression can be used to explain how the value observed for one measurement is related to the value observed for another quantity of interest. Generally, regression allows for a simple relationship between the different types of measurements, such that as the value of one measurement changes, then we would expect the other measurement to change proportionally. Regression allows data scientists to do a couple of useful things. Firstly, it enables them to interpret data, potentially providing the chance to understand the cause of the relationship behind the measurements of interest. For instance, a relationship between data related to smoking and cancer could be identified, which would help to identify that smoking increases the risk of cancer. Secondly, it allows for predictions of future measurements based on observing just some of the measurements. If we know how much someone smokes, we can use regression to predict their chances of getting cancer in the future. This is based on the data we have seen before from other people, including how much they smoked and whether they went on to develop cancer.

3-SECOND SAMPLE
Regression predicts values based on the data collected and is one of the most important tasks in data science.

3-MINUTE ANALYSIS
Regression is not always as simple as predicting one measurement from another. Sometimes there are millions of pieces of related data that need to go into the regression model, for example DNA data, and sometimes the different pieces of data have complex relationships with each other. More complex regression methods allow for situations such as this, but often require much more complex mathematics. A big part of data science is choosing the best regression model for the data available.

RELATED TOPICS
See also
DATA COLLECTION
page 16

REGRESSION TO THE MEAN
page 44

OVERFITTING
page 56

3-SECOND BIOGRAPHIES
CARL FRIEDRICH GAUSS
1777–1855
German mathematician who discovered the normal (Gaussian) distribution in 1809, a critical part of most regression methods.

FRANK E. HARRELL
fl. 2003–
Professor of Biostatistics at the Vanderbilt University, Nashville, and author of renowned textbook *Regression Modelling Strategies*.

30-SECOND TEXT
Vinny Davies

Regression helps data scientists understand relationships within collected data and make predictions about the future.

16 February 1822
Born in Birmingham,
England

1844
Receives BA degree,
Cambridge

1850–2
Travels in southwest
Africa (now Namibia)

1863
Creates the first modern
weather maps and
discovers the anti-
cyclone phenomenon

1869
Publishes *Hereditary
Genius*, a first study of
the inheritance of talent

1883
Coins the term 'eugenics'

1885
Discovers the
phenomenon of
regression to the mean,
leading eventually to
modern multivariate
analysis

1888
Discovers and names
correlation

1892
His book *Fingerprints*
launches a new era in
forensic science

17 January 1911
Dies in Surrey, England

FRANCIS GALTON

Francis Galton created the key to modern data analysis: the framework for the study of statistical association. Galton was born into a notable English family in 1822. Subsequently, however, the family would be best known for him and for his cousin Charles Darwin. Galton attended Cambridge, where he learned that formal mathematics was not for him, and while he then studied medicine, that profession, too, did not inspire him. When Galton was 22 years old his father died, leaving him sufficient wealth that he was able to live the rest of his life independent of need. For a few years he travelled, and for a nearly two-year period from 1851, went deep into southwest Africa, where he explored and met the people. At one point he helped negotiate a peace between two tribes.

In 1853 Galton married and settled down to a life in science. At first he wrote about travel, and he invented new forms of weather maps that incorporated glyphs showing details on wind, temperature and barometric readings. From these he discovered the anti-cyclone phenomenon, where a drop of barometric pressure reverses the cyclonic wind motion in the northern hemisphere. With the publication of his cousin Darwin's book *The Origin of Species* in 1859, Galton's main interest shifted to the study of heredity, anthropology and psychology. His most lasting inventions were the statistical methods he devised in those pursuits.

Galton invented correlation and discovered the phenomenon of regression, and he may, with some justice, be credited with taking the first major steps to a real multivariate analysis. His ideas are basic to all proper studies of statistical prediction, and to twentieth-century Bayesian analysis as well. Galton coined the term 'eugenics' and he promoted certain parts of this, but also wrote against others that would lead much later to associating eugenics with genocidal practices in the mid-twentieth century. Galton opposed the practice of creating heritable peerages and he encouraged granting citizenship to talented immigrants and their descendants. Some of his studies of inheritance came close to but did not reach Mendelian genetics, but he did help create the methods that would lead to the explosive development of biology after Mendel's work was rediscovered in 1901. Galton pioneered the use of fingerprints as a method of identification. He died childless in 1911, leaving his moderate fortune to endow a professorship and research at University College London

CLUSTERING

the 30-second data

Splitting data samples into

relevant groups is an important task in data science. When the true categories for collected data are known, then standard regression techniques – often called 'supervised learning' – can be used, to understand the relationship between data and associated categories. Sometimes, however, the true categories for collected data are unknown, in which case clustering techniques, or unsupervised learning, can be applied. In unsupervised learning, the aim is to group samples of data into related groups or clusters, usually based on the similarity between measurements. The meaning of these groups is then interpreted, or the groups are used to inform other decisions. A simple example of clustering would be to group animals into types based on characteristics. For instance, by knowing the number of legs/arms an animal has, a basic grouping can be created without knowing the specific type of animal. All the two-legged animals would likely be grouped together, and similarly animals with four and six legs. These groups could then easily be interpreted as birds, mammals and insects respectively, helping us learn more about our animals.

RELATED TOPICS
See also
LEARNING FROM DATA
page 20

REGRESSION
page 24

STATISTICS & MODELLING
page 30

3-SECOND SAMPLE
Sometimes data scientists don't have all the necessary data to carry out regression, but in many cases clustering can be used to extract structure from data.

3-MINUTE ANALYSIS
Netflix users aren't divided into specific categories, but some users have similar film tastes. Based on the films that users have watched or not watched, users can be clustered into groups based on the similarity of their watched/unwatched movies. While trying to interpret the meaning of these groups is difficult, the information can be used to make film recommendations. For instance, a user could be recommended to watch *Ironman* if they hadn't watched it but everyone in their cluster had.

3-SECOND BIOGRAPHIES
TREVOR HASTIE
1953–
Professor at Stanford University and co-author of *The Elements of Statistical Learning.*

WILMOT REED HASTING JR
1960–
Chairman and CEO of Netflix, who co-founded the company in 1997 as a DVD postage service.

TONY JEBARA
1974–
Director of Machine Learning at Netflix and Professor at Columbia University, USA.

30-SECOND TEXT
Vinny Davies

Clustering enables the grouping of data and the understanding of any connections.

STATISTICS & MODELLING

the 30-second data

When most people hear
'statistics' they think of a statistic, for example
a percentage. While statistics are an important
part of data science, more important is the
discipline of statistics, with statistical modelling
methods for regression and clustering being
some of the most used techniques in data
science. Statistics methods are known for
providing relatively simple and easily
interpretable techniques for analysing data.
For this reason, statistical methods are usually
the first place to start for most data scientists,
although they are often marketed as machine
learning, because it sounds cooler and the
difference between them can be unclear.
Statistical modelling methods often go well
beyond simple regression and clustering
techniques, but what they almost all have in
common is interpretability. Statistical models
are usually designed to clearly identify the
relationships between different measurements,
giving actionable results, which can guide
policy in areas like medicine and society.
This characteristic of statistical models is
vital for helping to work out whether a
relationship between measurements is
due to a coincidence or resulting from an
underlying causal relationship between the
sources of the measurements.

RELATED TOPICS
See also
REGRESSION
page 24

CLUSTERING
page 28

CORRELATION
page 42

3-SECOND BIOGRAPHIES
REVEREND THOMAS BAYES
1701–61
British statistician and
church minister famous for
formulating Bayes theorem,
the key to Bayesian statistics.

SIR DAVID COX
1924–
Prominent British statistician
and former president of the
Royal Statistical Society.

30-SECOND TEXT
Vinny Davies

3-SECOND SAMPLE
Statistics gives us many
of the basic elements of
data science, such as
percentages, but much
more interesting are
the numerous methods
provided by statistical
modelling.

3-MINUTE ANALYSIS
Bayesian statistics takes
prior information (data)
that is already known,
to help inform how to
analyse the same type
of future data. This
information can come in
many forms, including
measurements from a
related study or knowledge
about the range of
measurements that
could occur. In Bayesian
modelling, we can
incorporate this
information into our
model in a structured
and mathematical way,
enabling more informed
modelling choices, even
with a small amount
of data.

*Statistics is the
backbone of data
science and statistical
modelling methods are
always the first place
to start when looking
at new data.*

MACHINE LEARNING

the 30-second data

The idea of machine learning is to teach computers to learn and improve over time in an automated manner, without the need for human assistance. Algorithms can be implemented into systems, where they can make decisions automatically, often speeding up the decision-making process and reducing the possibility of human error. Within the system, the machine learning algorithms use the data they receive to make predictions about the future, helping the system to operate and choose between different options. The algorithm then updates itself based on what it learned from the information it received, ensuring that it continues to make optimal decisions in the future. An everyday example of machine learning in action is Spotify. The music app has millions of users, and data all about the type of music those users like, based on the songs they have listened to. When a new user joins, Spotify knows very little about them and will recommend songs almost at random. But as the user listens to songs, the algorithm continually learns about the music preferences of that user and how they relate to the preferences of other users. The more songs the user listens to, the better the algorithm becomes, and the song recommendations improve for that user.

RELATED TOPICS
See also
NEURAL NETWORKS &
DEEP LEARNING
page 34

PRIVACY
page 86

ARTIFICIAL INTELLIGENCE (AI)
page 148

3-SECOND BIOGRAPHIES
YANN LECUN
1960–
Professor at New York University and Chief AI Scientist at Facebook.

ANDREW NG
1976–
Stanford University Professor famous for his work in machine learning, as well as for founding the Google Brain project and online learning platform Coursera.

30-SECOND TEXT
Vinny Davies

3-SECOND SAMPLE
Machine learning gives us the ability to learn from data without human intervention, allowing us to automate tasks and remove human decision making.

3-MINUTE ANALYSIS
Films like *The Terminator* can make machine learning seem scary. So how far are we from robots stealing our jobs and Skynet taking over the world? While machine learning may take over some small jobs, it is unlikely that a truly intelligent computer will be designed which would take all our jobs. Even if that happened, humans would still need to supervise the computer to ensure it didn't make any inhumane decisions (or create robot Arnold Schwarzeneggers!).

The more data that is collected, the more a machine will learn, and the smarter it will become.

NEURAL NETWORKS & DEEP LEARNING

the 30-second data

Originally inspired by the human brain, neural networks are one of the most common machine learning methods. Like the brain, neural networks consist of a network of interconnected (artificial) neurons which allow the interpretation of images or other types of data. Neural networks are used to help in everyday life, from finding the faces in smartphone photos, to reading addresses on envelopes, ensuring they go to the correct location. Deep learning is a group of methods based around neural networks, but with a much larger number of layers of interconnecting artificial neurons. One of the uses of deep learning is analysing and responding to messages, either in the form of text (customer service chat bots for example) or speech (such as Alexa or Siri). However, the biggest use of deep learning is in image processing. Deep learning can be used to analyse the images captured by driverless cars, interpreting the results and advising the car to adjust its course as needed. It is also beginning to be applied in medicine, with its ability to analyse images such as MRIs or X-rays, making it a good way of identifying abnormalities, such as tumours.

RELATED TOPICS

See also
MACHINE LEARNING
page 32

IBM WATSON & GOOGLE'S DEEPMIND
page 94

ARTIFICIAL INTELLIGENCE (AI)
page 148

3-SECOND BIOGRAPHIES
FRANK ROSENBLATT
1928–71
American psychologist famous for developing the first method that resembles a modern-day neural network.

YOSHUA BENGIO
1964–
Canadian computer scientist famous for his work on neural networks and deep learning.

30-SECOND TEXT
Vinny Davies

3-SECOND SAMPLE
Many modern technologies rely on neural networks and deep learning, which have given us driverless cars and virtual assistants.

3-MINUTE ANALYSIS
Amazon has created a supermarket where you don't need to scan items. You just pick up items, put them in your bag and walk out. The supermarket works by videoing everyone as they shop and using deep learning to identify each item customers pick up, noting whether they put it in their bag or back on the shelf. When you walk out, the cost of your items is simply charged to your account.

While deep learning is a highly sophisticated process, its prevalence in the future will depend on the level of trust it can garner.

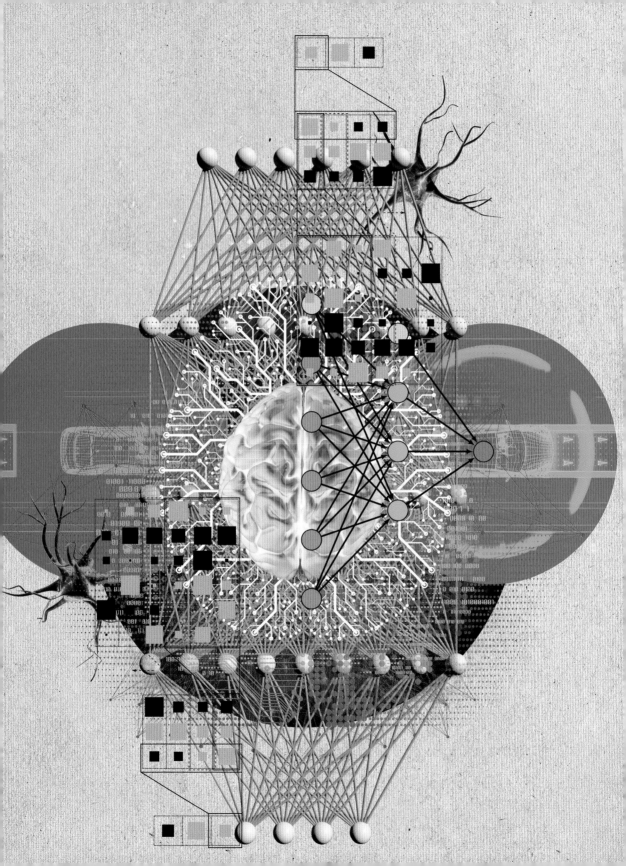

UNCERTAINTY ◑

algorithmic bias Algorithms learn how to make decisions by processing examples of humans performing the same task. If this data is taken from a prejudiced source, the model will learn to replicate those prejudices.

automated system Repetitive tasks or calculations carried out by computers, e.g. automated passport gates at airports, self-driving cars and speech-to-text software.

causation If a change in one variable directly causes a change in another variable, causation exists.

correlation Two variables are correlated if a change in one is associated with a change in the other.

cross-validation Fitting and testing a predictive model on different subsets of a data set. Cross-validation can be a method for fine-tuning a model's parameters, and can also provide better estimates of a model's performance.

data point A piece of information. A single data point may consist of several quantities or variables, provided they are all associated with a single observation from the real world.

Gallup poll A series of regular surveys, conducted by the company Gallup, to gauge public opinion on a range of political, economic and social issues.

natural variation Changes or fluctuations that occur in populations or the natural world over time, e.g. natural variations in a country's birth rate over time.

noise Random variations in data collected or measured from the real world. Minimizing or accounting for the effects of noise in data is a crucial step in many statistical analyses.

non-probability sampling Method of sampling from a population, where not all members of the population have an equal chance of selection.

non-response bias Introduced when people who are able or willing to respond to a survey differ significantly from people who do not or cannot respond.

null hypothesis The hypothesis that there is no significant difference between populations, meaning that any observed difference is due to error, noise or natural variation.

p-value The probability that the results observed in an experiment would occur if the null hypothesis was true.

predictive model A mathematical model which predicts the value of an output, given values of an input.

regularization A technique to discourage overfitting in models.

sample A subset of a population, selected for participation in a study, experiment or analysis.

sampling Selecting members of a population as participants in a study or analysis.

selection bias Introduced when samples for a study are selected in a way that does not result in a representative sample.

self-selection bias Introduced when participants assign themselves to a study, or a group within a study. This may lead to a sample that is biased and unrepresentative of the population.

statistically significant A result that is very unlikely to have occurred if the null hypothesis were true. For example, if a study

was investigating whether students who drink coffee perform better in exams than students who don't, the null hypothesis would be 'there is no difference in exam performance between students who do and don't drink coffee.' If a study found significant differences in performance between coffee drinking and non-coffee drinking students, the null hypothesis could be rejected.

time series analysis The analysis of a signal or variable that changes over time. This can include identifying seasonal trends or patterns in the data, or forecasting future values of the variable.

training data Many machine learning models are fitted to training data, which consists of inputs and their corresponding outputs. The model 'learns' the relationship between the inputs and outputs, and is then able to predict the output value for a new, unseen input value.

univariate and multivariate time-dependent data Univariate time-dependent data consists of the values of a single variable over time, whereas multivariate time-dependent data consists of the values of more than one variable.

SAMPLING

the 30-second data

'Garbage in, garbage out': data scientists know that the quality of their data determines the quality of their results, so most of them have learned to pay careful attention to measurement collection. When analysts can work with an entire population's data – such as Netflix tracking the film-watching habits of its subscribers – drawing conclusions can be a straightforward matter of just crunching numbers. But that completeness is not always practical. In criminal healthcare fraud investigations, the 'full population' would be health claims records numbering in the trillions. Instead, lawyers might have data scientists strategically choose a subset of records from which to draw conclusions. Other times, as with political polling, all that is available is a sample. If the sample is a randomly chosen one, statistical theories exist to tell us how confident we should be in our generalizations from sample to population. Increasingly, data scientists are relying on what is known as 'non-probability sampling', where the sample is not chosen according to any randomization scheme. So using Twitter data to track the buzz of a candidate or brand will not give a random sample representing the entire population – but it still has worth.

RELATED TOPICS

See also

DATA COLLECTION
page 16

SAMPLING BIAS
page 48

VOTE SCIENCE
page 90

3-SECOND SAMPLE
When the entire population of interest can't be measured or questioned, a sample is taken – but how that is done is as much an art as it is a science.

3-MINUTE ANALYSIS
In 1936, the US was in the Great Depression, and a conservative small-town mayor was challenging President Roosevelt for office. The most influential magazine of the time, *Literary Digest*, polling 2.4 million voters, predicted a challenger's landslide. Wrong: Roosevelt swept the nation. What happened? The sample was large but biased; the magazine polled its subscribers – car owners and telephone users – all wealthier than average. Within two years *Literary Digest* had folded, and a new science of statistical sampling was launched.

3-SECOND BIOGRAPHIES

ANDERS NICOLAI KIÆR
1838–1919
First to propose that a representative sample be used rather than surveying every member of a population.

W. EDWARDS DEMING
1900–93
Wrote one of the first books on survey sampling, in 1950, which is still in print.

GEORGE HORACE GALLUP
1901–84
American pioneer of survey sampling techniques and inventor of the Gallup poll.

30-SECOND TEXT
Regina Nuzzo

Statisticians work to find out the accuracy of conclusions even from irregular samples.

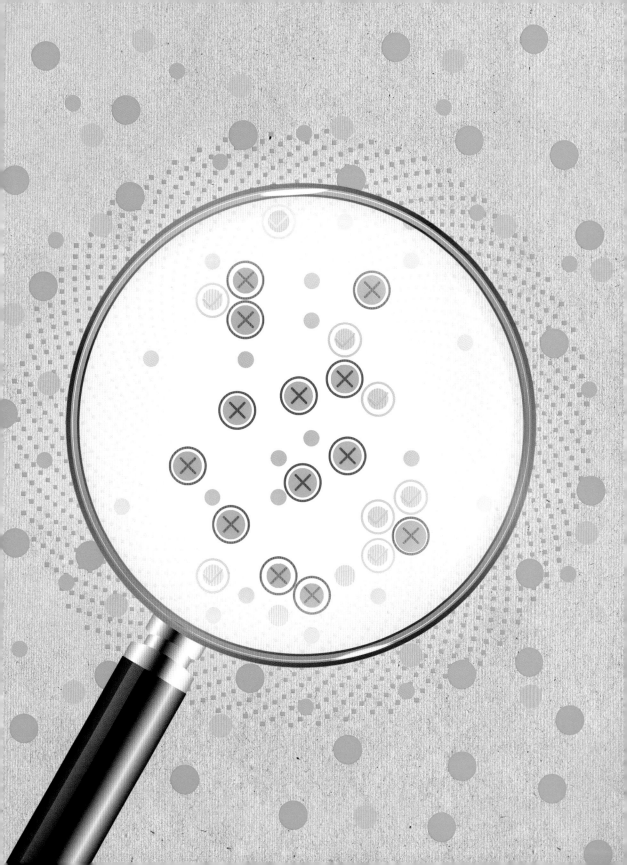

CORRELATION
the 30-second data

A correlation is a kind of dance –
a 'co-relation' – between two features in a data
set. A positive correlation means the dancers
are moving more or less in the same direction
together: when crude oil prices rise, for
example, retail petrol prices also tend to
rise. A negative correlation means the dancers
are still in sync but are moving in opposite
directions: longer website loading times are
associated with lower customer purchase
rates. Correlations can only capture linear
relationships, where two features can be
visualized on a graph together as a straight
line. That means an analysis of business
characteristics such as staff cheerfulness and
customer satisfaction might return a 'zero
correlation' result, hiding a more interesting
story underneath: a curvilinear relationship,
where customers dislike too little cheerfulness
but also too much. Another problem is that
correlation is not the same as causation. Sales
of ice cream and drowning deaths are positively
correlated, but of course that does not mean
that banning the sale of ice cream will save
lives. The causation culprit is often a third
characteristic (daily temperature). It is up to
the analyst to intelligently use all available
information to figure out whether the
apparent cause-and-effect is real.

RELATED TOPICS
See also
REGRESSION TO THE MEAN
page 44

OVERFITTING
page 56

3-SECOND BIOGRAPHIES
KARL PEARSON
1857–1936
English mathematician
who developed Pearson's
correlation coefficient, the
most common way to measure
correlation.

JUDEA PEARL
1936–
Israeli-American computer
scientist and philosopher
whose work has helped
researchers distinguish
correlation from causation.

30-SECOND TEXT
Regina Nuzzo

3-SECOND SAMPLE
At the heart of modern
data science lies a
surprisingly simple
concept: how much do
two things move in sync
with each other?

3-MINUTE ANALYSIS
In 2014, for a fun project
before final exam week,
Harvard law student Tyler
Vigen purposely set out to
find as many coincidental
correlations as possible
across multiple data sets.
His website Spurious
Correlations quickly went
viral, allowing millions of
visitors to view graphs
showing the high
correlation over time
between oddball variable
pairs, such as the number
of people who have died by
becoming tangled in their
bedsheets and the per
capita cheese consumption
in the US.

*Graphs illustrating
dynamic relationships
can be a data scientist's
most powerful tool.*

REGRESSION TO THE MEAN

the 30-second data

Can stats explain the strange phenomenon where top rookie athletes fall from glory and go on to a disappointing second season? The usual explanation is that stars hit this slump because they choke under pressure and attention from a stellar debut. But data whizzes know better – it is just a statistical affair called regression to the mean. And it's not unique to sports; you can find examples everywhere. Why do the most intelligent women tend to marry men less intelligent than themselves? Why was a company's surprisingly profitable quarter immediately followed by a downturn? Why do hospital emergency departments get slammed the moment someone remarks, 'Wow, it's quiet today'? It is probably not a cause-and-effect story (or superstitious jinx). Regression to the mean says that extreme events don't stay extreme forever; they tend back towards the average, just on their own. It is not that any true effect in the data disappears – to the contrary, native athletic talent persists, good fiscal management carries on – but the extreme luck that pushed an individual into the top tiers today is likely to fade out tomorrow. Data scientists know to be on guard for this effect, lest they be fooled into spotlighting trends that aren't real.

RELATED TOPICS
See also
REGRESSION
page 24

CORRELATION
page 42

3-SECOND BIOGRAPHIES
FRANCIS GALTON
1822–1911
First coined the concept of regression to the mean in his study of genetics and height.

DANIEL KAHNEMAN
1934–
Nobel Laureate who suggested regression to the mean might explain why punishment seems to improve performance.

30-SECOND TEXT
Regina Nuzzo

3-SECOND SAMPLE
'What goes up must come down' – it may seem obvious, but in stats this is easy to miss, and it can lead to some puzzling trends.

3-MINUTE ANALYSIS
Regression to the mean is especially important when analysing data that has been chosen based on a measurement that has exceeded some threshold – for example, patients whose last blood pressure measurement was considered dangerous, or patients with a sudden worsening of depression symptoms. In fact, about a quarter of patients with acute depression get better no matter what – with drugs, therapy, placebo or nothing at all – leading some researchers to question the usefulness of standard depression treatments.

Stats can help explain dramatic swings of fortune in sports, as well as in life.

CONFIDENCE INTERVALS

the 30-second data

3-SECOND SAMPLE
Confidence intervals are almost magical in their ability to take a piece of limited information and extrapolate it to the entire population.

3-MINUTE ANALYSIS
Beware journalists reporting numbers without confidence intervals. For example, a 2017 *Sunday Times* article highlighted a reported drop of 56,000 employed people in the UK, saying 'it may signal the start of a significantly weaker trend'. Digging deeper into Office for National Statistics reports, however, reveals a confidence interval for the true change in number employed running from a 202,000 decline to a 90,000 increase. So employment may not have dropped at all – it might have actually improved!

When you're lucky enough to get data on an entire population – all customer purchases from a web vendor last year, say – then getting the true average is easy: just crunch the numbers. But when all you get is a sample of the population – like satisfaction ratings from only 1,000 customers out of 1 million – knowing the true average value is much trickier. You can calculate the average satisfaction rating of your sample, but that's just a summary of these particular 1,000 customers. If you had taken another random 1,000 customers, you would get a different average. So how can we ever talk about the average satisfaction of all million people? That is where confidence intervals come to the rescue – one of the tools statisticians use in pursuit of their ultimate goal of drawing conclusions about the world based on limited information. Statisticians have worked out ingenious maths that takes information from one sample and uses it to come up with a whole range of plausible values for the average of the entire population. So instead of just saying the average satisfaction rating in one sample was 86 per cent, you can say, with some confidence, that the average satisfaction in the entire customer population is between 84 and 88 per cent – which is much more valuable information.

RELATED TOPICS
See also
SAMPLING
page 40

STATISTICAL SIGNIFICANCE
page 54

3-SECOND BIOGRAPHY
JERZY NEYMAN
1894–1981
Polish mathematician and statistician who introduced confidence intervals in a paper published in 1937.

30-SECOND TEXT
Regina Nuzzo

Making conclusions about the big picture with confidence is where the field of statistics shines.

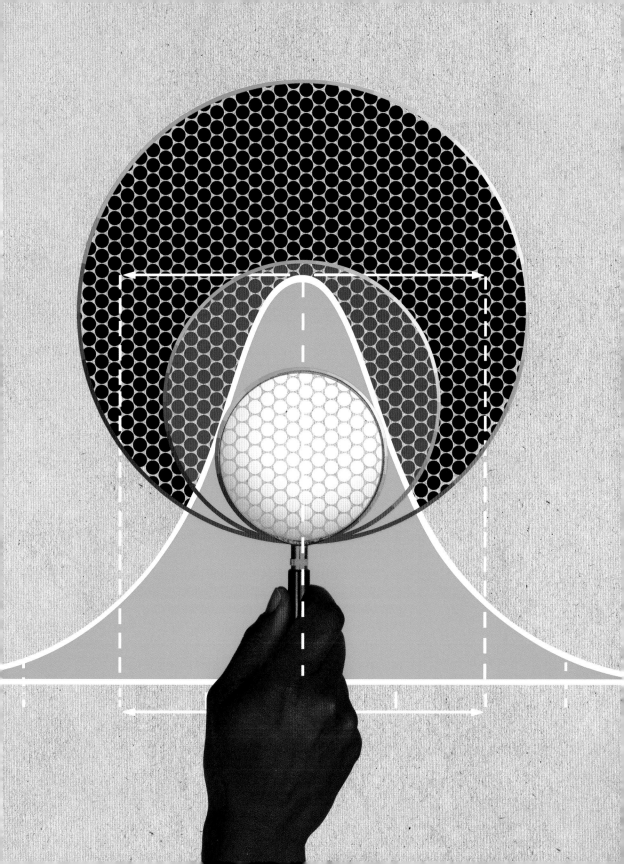

SAMPLING BIAS

the 30-second data

Data points are like gold nuggets, so data scientists eagerly scoop up whatever they can find. Smart analysts do something even more valuable: they stop, look around and ask what happened to all the nuggets that aren't lying around in plain sight. Are those left-out data different in any systematic way from the data that were easy to collect? Take, for example, a report's estimate that 10 per cent of men suffer from impotence – results that were based on a survey of patients at an andrology health clinic. This selection bias happens when the participants chosen differ in important ways from the ones not chosen (such as, here, their sexual health). Related to this is self-selection bias, where, for example, service satisfaction ratings can easily skew negatively if only the most irate customers take time to respond. Likewise, there is non-response bias; medical studies, for example, can be misleading if researchers ignore the fact that those participants most likely to drop out are also the ones who are the sickest. Sometimes it is possible to statistically correct for a bias problem, but recognizing the problem in the first place is often the hardest part.

RELATED TOPICS
See also
DATA COLLECTION
page 16

SAMPLING
page 40

OVERFITTING
page 56

3-SECOND BIOGRAPHIES
ABRAHAM WALD
1902–50
Hungarian mathematician whose work on Second World War aircraft damage illustrates the concept of survivorship bias.

CORINNA CORTES
1961–
Danish computer scientist and Head of Google Research, who works on sample bias correction theory.

30-SECOND TEXT
Regina Nuzzo

3-SECOND SAMPLE
It's almost a paradox in data science: what's not in a data set can be even more important than what's in it.

3-MINUTE ANALYSIS
In the Second World War the American military gathered data on bullet holes from planes returned from European battles. Where were the highest bullet densities, they asked, so extra armour could be added to spots where planes are shot at the most? Statistician Abraham Wald turned the question on its head. These data only show where planes that managed to make it back home had been hit, he pointed out. Planes were getting shot at in other places, but these planes, hit in other spots, didn't survive. So the armour belonged, he said, where the bullet holes weren't.

A skilled data scientist will seek out gaps in the data collection process and analyse their potential impact.

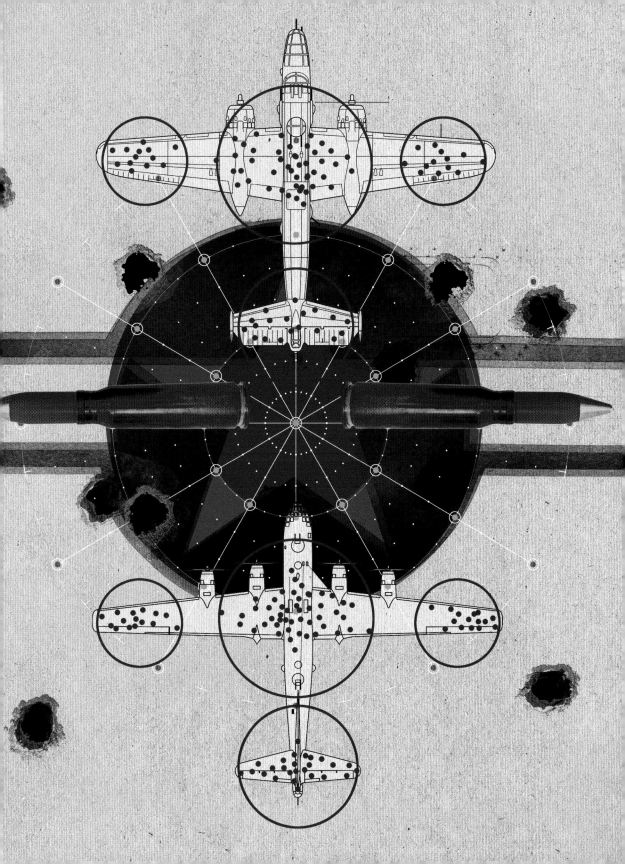

BIAS IN ALGORITHMS

the 30-second data

3-SECOND SAMPLE

Can a computer be racist, sexist or homophobic? Human biases are often built into automated systems, with serious consequences for the most vulnerable groups in society.

3-MINUTE ANALYSIS

As many machine learning models are developed by private companies, their training data and source code are not open to scrutiny. This poses challenges for journalists investigating algorithmic bias. In 2016, an investigation by the news outlet ProPublica used Freedom of Information requests to reverse-engineer the COMPAS algorithm, used in the US to predict the likelihood of criminals reoffending. They uncovered racial discrimination, raising questions on regulation and transparency in AI.

Algorithms learn how to make decisions by processing examples of humans performing the same task. An algorithm for sentencing criminals might be trained on thousands of historic decisions made by judges, together with information about the offenders and their crimes. If this training data is taken from judges who give harsher sentences to people of colour, the model will learn to replicate those prejudices. In 2018, the Massachusetts Institute of Technology's (MIT) Media Lab showed that face recognition systems developed by Microsoft, IBM and China's Face++ were all significantly worse at detecting female faces, and performed poorly on images of darker-skinned women. With police forces in the UK and US testing automated facial recognition systems for crime prevention, low accuracies and false alarms could have far-reaching consequences for civil liberties. In 2018 Amazon scrapped an automated CV screening tool due to gender bias. The system was trained on data from previous successful candidates, who were mostly male, due to existing imbalances in the technology industry. This produced a tool that penalized applications containing phrases more likely to appear in women's résumés, such as 'women's football team'. The algorithm learned to equate men's CVs with success, and women's with failure.

RELATED TOPICS

See also
SAMPLING BIAS
page 48

ARTIFICIAL INTELLIGENCE (AI)
page 148

REGULATION
page 150

3-SECOND BIOGRAPHY

JOY BUOLAMWINI
fl. 2011–
Computer scientist and digital activist, based at the MIT Media Lab, and founder of the Algorithmic Justice League.

30-SECOND TEXT

Maryam Ahmed

The potential for bias might sound far-fetched, but algorithm bias poses a very real problem requiring creative solutions.

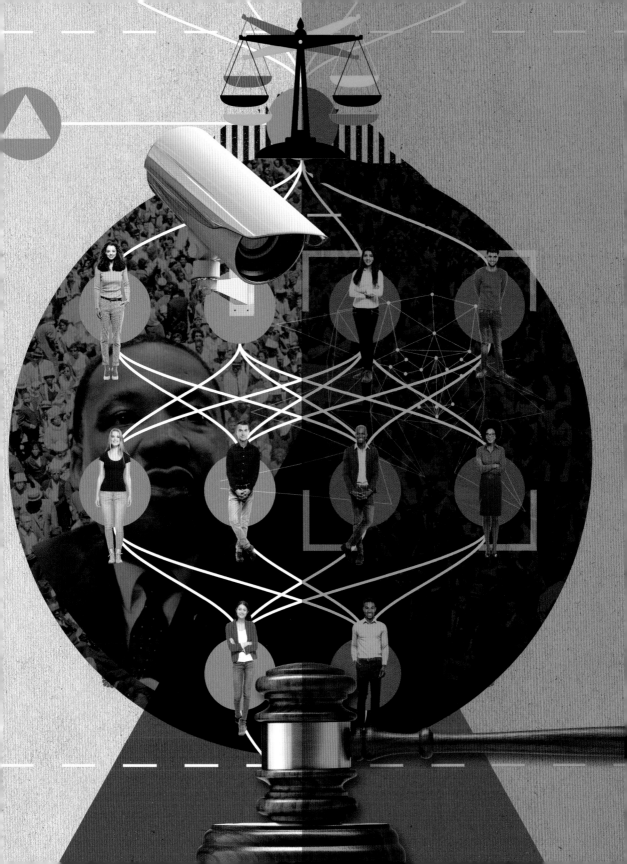

18 October 1919
Born in Kent, England

1953
Receives his PhD at
University College
London

1959
Marries Joan Fisher; he
later gives her statistical
advice as she writes her
1978 biography of her
father, Ronald A. Fisher

1960
Moves to Madison,
Wisconsin, to start a new
Department of Statistics

1970
Publishes *Time Series
Analysis* (with Gwilym
Jenkins). In subsequent
years he also develops
forecasting methods,
based upon difference
equation methods, with
other authors

1973
Published *Bayesian
Inference in Statistical
Analysis* (with George C.
Tiao)

1978–9
Serves as President of
the American Statistical
Association, and of the
Institute of Mathematical
Statistics

1985
Elected Fellow of the
Royal Society of London

28 March 2013
Dies in Madison,
Wisconsin, USA

GEORGE BOX

George Box was born in England in 1919. He had studied chemistry before being called up for service during the Second World War, and he gained his first introduction to statistics when, while engaged in war work, he encountered problems with the interpretation of experimental data. Someone suggested he visit British statistician and geneticist Ronald A. Fisher, at the time working from home because his laboratory at Cambridge had been closed for the duration of the war. The visit opened Box's eyes to the world of 'data science' (a then unknown term), and after the war he went to University College London for graduate study. There, as later in life, he plotted his own course, concentrating on understanding the role of statistics in scientific and engineering investigation.

Box's early work was as a statistician at Imperial Chemical Industries, where he was involved with the design of experiments. In one early paper he introduced the word and concept of 'robustness' to statistics: the idea that the validity of some ('robust') statistical procedures could withstand even large departures from conditions thought to be key to their use. After a few years that included time in Princeton (where he met and married Joan Fisher, one of Ronald's daughters), Box moved in 1960 to start a new department of statistics at the University of Wisconsin in Madison, where he spent the rest of his life and did his most influential work.

Box was a great catalyst in collaborative scientific investigations. He ran a famous evening 'Beer seminar' weekly, where a scientist would briefly present a problem and the assembled group would produce innovative solutions, some with great lasting effect. With various co-authors he created new methods of time series analysis for univariate and multivariate time-dependent data, new ideas for the use of Bayesian methods and new approaches to experimental design, including 'evolutionary operation', an approach that permitted bringing experiments to the manufacturing floor and involving line workers in continuously improving processes without interrupting production. He was a great advocate for keeping the scientific question always at the forefront, and for the importance of good experimental design. He employed mathematical models, but famously is quoted as cautioning that 'all models are wrong, but some are useful'. He died in Madison in 2013.

Stephen Stigler

STATISTICAL SIGNIFICANCE

the 30-second data

RELATED TOPICS
See also
STATISTICS & MODELLING
page 30

SAMPLING
page 40

3-SECOND SAMPLE
Are those interesting patterns in a data set just a random fluke? A century-old stats tool can help answer that.

3-MINUTE ANALYSIS
P-values are easy to hack. In 2015, media around the world excitedly reported on a study showing that chocolate leads to weight loss. Then the author revealed the truth: he was a journalist, the data was random and his results just a false-positive fluke. He knew that in 5 per cent of studies p-values will be smaller than 0.05 just by chance. So he ran 18 separate analyses of random data – and then reported only the deliciously statistically significant one.

It is worth getting to know the p-value, because this tiny number boasts outsized importance when it comes to drawing conclusions from data. The tininess is literal: a p-value is a decimal number between 0 and 1. It is calculated when you have a question about the world but only limited data to answer it. Usually that question is something like, 'Is there something real happening here in the world, or are these results just a random fluke?' If you toss a coin 100 times and it comes up heads every time, you might suspect that the coin is double-headed, but there is still the possibility (however negligible) that the coin is fair. The p-value helps support your scepticism that this event didn't happen by accident. By tradition, results with a p-value smaller than 0.05 get labelled 'statistically significant' (in the case of the coin, getting all heads from five flips). It is this label that people often use for reassurance when making decisions. But there is nothing magical about the 0.05 threshold, and some experts are encouraging researchers to abandon statistical significance altogether and evaluate each p-value on its own sliding scale.

3-SECOND BIOGRAPHIES
KARL PEARSON
1857–1936
British statistician who first formally introduced the p-value.

SIR RONALD FISHER
1890–1962
British statistician who popularized the p-value in his 1925 book for researchers.

30-SECOND TEXT
Regina Nuzzo

P-values help statisticians work out whether results are a random fluke – or not: the gold standard of statistical evidence has some major flaws.

OVERFITTING

the 30-second data

Building a predictive model

involves finding a function that describes the relationship between some input and an output. For example, a data scientist may want to predict a university student's final grade based on their attendance rate in lectures. They would do this by fitting a function to a 'training' set of thousands of data points, where each point represents a single student's attendance and grade. A good model will capture the underlying relationship between grades and attendance, and not the 'noise', or natural variation, in the data. In this simple example, a reliable model may be a linear relationship. When a new student's attendance is added, the model will use it to predict their final grade because it generalizes to the student population as a whole. An overfitted model will involve more parameters than necessary; instead of fitting a straight line to the university data set, an overenthusiastic data scientist might use a very complex model to perfectly fit a contorted, meandering curve to the training data. This will not generalize well, and will perform poorly when presented with data for a new student. Understanding that a complex model is not always better is a crucial part of responsible and thoughtful data science practice.

3-SECOND SAMPLE
Beware of complex models that fit the data perfectly. It is likely they are overfitted, and will predict poorly when presented with new data points.

3-MINUTE ANALYSIS
There are ways to avoid overfitting. Cross-validation gives an estimate of how well a model will work in practice, by training the model on a subset of the training data and testing its performance on the remaining subset. Regularization is a technique that penalizes a model for being too complex; in the university example, a line would be preferred over a curve.

RELATED TOPICS
See also
REGRESSION
page 24

STATISTICS & MODELLING
page 30

MACHINE LEARNING
page 32

30-SECOND TEXT
Maryam Ahmed

If a model's performance seems too good to be true, then it probably is!

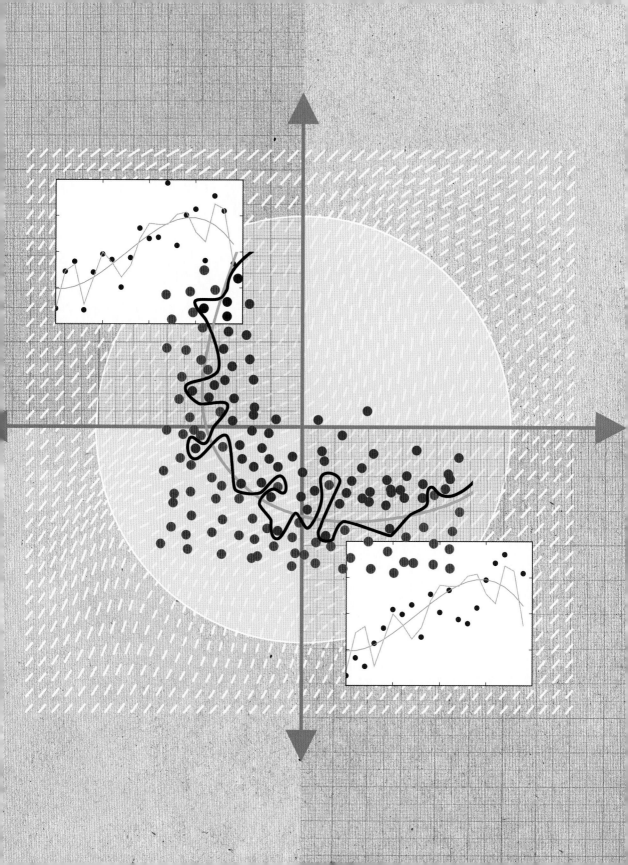

SCIENCE

SCIENCE
GLOSSARY

anthropogenic Event or phenomenon that is caused by humans, for example climate change.

blind analysis Conducted where researchers cannot see the correct measurements or answers; aims to minimize bias.

causal relationship If a change in one variable directly causes a change in another variable, a causal relationship exists between them.

correlation Two variables are correlated if a change in one is associated with a change in the other. A positive correlation exists if one variable increases as the other increases, or if one variable decreases as the other decreases. A negative correlation exists if one variable increases as the other decreases.

data set A set of information stored in a structured and standardized format; might contain numbers, text, images or videos.

debugging Finding and correcting errors in computer code.

diagnostics Identification of problems, typically human diseases or health conditions; also refers to the identification of computer bugs.

DNA The genetic code that governs the development, characteristics and functioning of every living organism. DNA is usually found in the nucleus of a cell, and consists of two long chains of building blocks called 'nucleotides', arranged in a double helix shape. In most humans, an individual's genome, or genetic code, is unique. Recent advances in genetic engineering have enabled the insertion, deletion and modification of genetic material in DNA.

epidemiological evidence Correlation between exposure to a risk factor, such as smoking, and incidence of a disease, such as lung cancer.

experimental design The process of designing robust studies and experiments, to ensure that any conclusions drawn from the results are reliable and statistically significant. This includes careful selection of experimental subjects to avoid sampling bias, deciding on a sample size, and choosing suitable methods for analysing results.

gene editing Process of editing the genome of a living organism by inserting, removing or modifying its DNA.

genome Genetic material, or chromosomes, present in a particular organism. The human genome consists of 23 pairs of chromosomes.

greenhouse gas A gas in the atmosphere which absorbs and radiates energy, contributing to the warming of Earth's surface. This causes the so-called 'greenhouse effect', which is necessary for supporting life on Earth. Human activity has led to an increase in greenhouse gases in the atmosphere, which have amplified the greenhouse effect and contributed to global warming. Greenhouse gases include water vapour, carbon dioxide and methane.

independent replication Validation of a study or experiment by independent researchers. This is done by repeating the procedure followed by the original researchers, to ensure the results can be replicated.

randomized trials Experimental design where participants or subjects are randomly allocated to treatment groups. For example, participants in a randomized drug trial could be randomly allocated to a group where they would either receive a placebo or a drug.

trendlines A way of visualizing the overall direction, or trend, of a variable over time. There are different methods for calculating trendlines, including a moving average, or a line of best fit calculated through linear regression.

CERN & THE HIGGS BOSON

the 30-second data

In 1964, Peter Higgs, Francois Englert, Gerald Guralnik, C.R. Hagen and Tom Kibble proposed the Higgs Mechanism to explain how mass was created in the universe. But evidence of the mechanism lay in the (elusive) discovery of an essential particle, dubbed 'Higgs boson', from which other fundamental particles derived their mass. By blasting particles into each other at incredibly high energies and then gathering data on the number of emergent particles as a function of particle energy, scientists hoped to identify spikes (in collisions at particular energy levels), which in turn would point to the creation of a particle, such as the Higgs boson, with that energy. Enter CERN, the world-famous European laboratory. Here, scientists built the Large Hadron Collider (LHC). Even in its infancy (2008), LHC's enormous capability was stunning: it was able to accelerate particles to about 14 billion times their energy at rest. By 2011, CERN had collected enough data – over 500 trillion collision events – for analysis. Not long after, several independent groups caught an energy spike in the very field where the Higgs was predicted to lie. This discovery was soon acknowledged by the scientific community, and both Higgs and Englert won acclaim as joint recipients of the 2013 Nobel Prize for Physics.

3-SECOND SAMPLE
CERN, a laboratory in Switzerland, is synergy of multinational proportions: here, top scientists convene to inspect and decode the constituents of matter via particle colliders, i.e. how the universe works.

3-MINUTE ANALYSIS
The LHC in CERN is connected to four separate detectors into which highly accelerated particles can be slammed. For the Higgs boson experiments, two detectors, ATLAS and CMS, were used. The fact that the same results were observed on both detectors lent significant credibility to the Higgs discovery, once again emphasizing the importance of independent replication in data analysis.

RELATED TOPIC
See also
MACHINE LEARNING
page 32

3-SECOND BIOGRAPHIES
PETER HIGGS
1929–
First proposed the Higgs Mechanism.

FRANCOIS ENGLERT
1932–
Also proposed the Higgs mechanism, independently of Higgs.

30-SECOND TEXT
Aditya Ranganathan

The reach of data science knows no bounds, being applied to explain the very workings of the universe.

ASTROPHYSICS
the 30-second data

3-SECOND SAMPLE
Photons from stars billions of light years away strike Earth, furnishing telescopes with eons-old galactic images – masses of data awaiting analysis.

3-MINUTE ANALYSIS
A major problem in data analysis is the tendency to interpret results as confirmations of pre-existing beliefs, which leads to furious debugging when outcomes clash with expectations and to slackening of error-detection when the two correspond. To decontaminate the debugging, physicists developed blind analysis, wherein all analysis happens before the final experimental results are revealed to the researcher. Blind analysis has gained popularity in areas of physics and may be making a foray into other fields such as psychology.

Astrophysics has become a big user and supplier of data science expertise. Most cosmology experiments involve scanning large amounts of data to make measurements that can only be statistically derived. The data is also searched for rare events. These statistical insights, in turn, elucidate the past – and future – of our universe. One example of a rare cosmological event is the production of a supernova – a star that explodes during its demise. Supernovae were used in the discovery of the accelerating expansion of the universe, for which Saul Perlmutter, Brian Schmidt and Adam Reiss won the 2011 Nobel Prize. The discovery hinged on automatically searching the sky for supernovae and collecting enough measurements of supernova brightness and redshift (a measure of how much photons have been stretched) in order to make statistically acceptable conclusions about trendlines. Supernovae have homogenous brightness, and it is this brightness that indicates how far a supernova is from a telescope, and how long light takes to reach us from that supernova; if light from older supernovae stretched less than from new supernovae, the universe must be stretching more now than before, implying that over time, the universe will continue to stretch ever more rapidly.

RELATED TOPIC
See also
CERN & THE HIGGS BOSON
page 62

3-SECOND BIOGRAPHIES
EDWIN HUBBLE
1889–1953
American astronomer who discovered the original expansion of the universe.

SAUL PERLMUTTER
1959–
American astrophysicist and Director of the Berkeley Institute for Data Science who won the 2011 Nobel Prize in Physics for the Accelerating Expansion of the Universe.

30-SECOND TEXT
Aditya Ranganathan

Data-driven measurements and experiments highlight the importance of data science to cosmology, and vice versa.

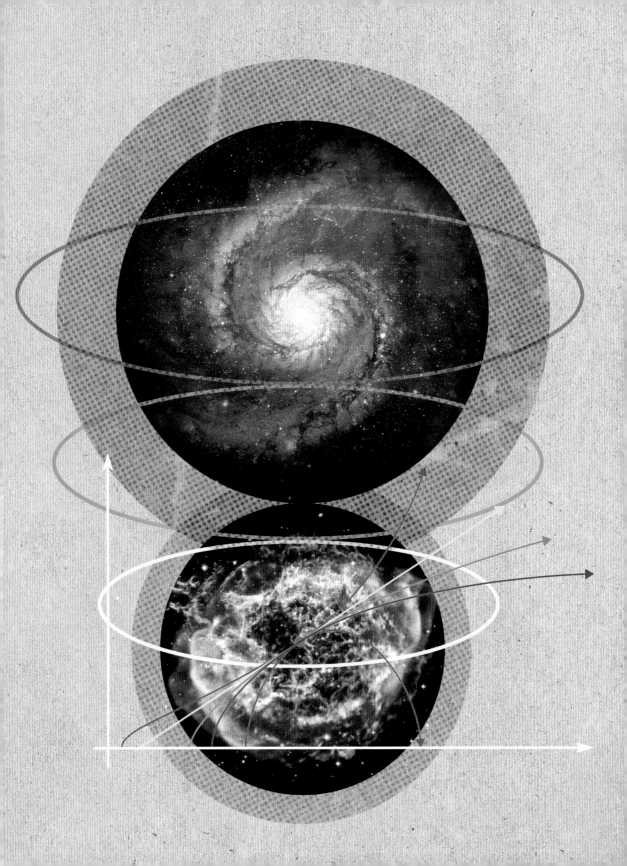

CRISPR & DATA

the 30-second data

Scientists are harnessing the power of a gene-editing tool called CRISPR that has revolutionized labs around the world. The precision engineering tool allows scientists to chop and change DNA in a cell's genetic code and could one day correct mutations behind devastating diseases such as Huntington's, cystic fibrosis and some cancers. CRISPR works like a pair of molecular scissors and cuts DNA at target genes to allow scientists to make changes to the genome. This technique has been used by scientists in the lab to make embryos resistant to HIV and remove genes that cause sickle-cell disease. But these molecular scissors are not perfect. One misplaced cut could cause irreparable damage that is passed on through generations. To make CRISPR more accurate, scientists are leveraging huge data sets generated from mapping the human genome. Researchers have used CRISPR to edit tens of thousands of different pieces of DNA and analysed the resulting sequences. From the data, scientists are developing machine learning algorithms that predict the exact mutations CRISPR can introduce to a cell, helping scientists to reduce any miscuts to the code of life.

RELATED TOPICS
See also
THE MILLION GENOME PROJECT
page 68

CURING CANCER
page 74

ETHICS
page 152

3-SECOND SAMPLE
Editing the human genome conjures images of science fiction, but it could be closer to reality thanks to the data science that is helping researchers to correct nature's mistakes.

3-MINUTE ANALYSIS
There is trepidation about where CRISPR technology will take science, and gene-editing of human embryos has raised ethical concerns – specifically around the possibility of introducing heritable alterations to the human genome. Some genome damage could go unnoticed and lead to unforeseen health issues, such as premature death or other genetic diseases. It is no surprise that CRISPR has sparked international debate on how it should be regulated.

3-SECOND BIOGRAPHIES
FRANCISCO J.M. MOJICA
1963–
One of the first researchers to characterize CRISPR and coin the acronym.

JENNIFER ANNE DOUDNA
1964–
Along with Emmanuelle Charpentier, proposed CRISPR as a gene-editing tool.

FRANK STEPHENS
1982–
Advocate for Down syndrome; Special Olympics athlete.

30-SECOND TEXT
Stephanie McClellan

Big data sets are helping to refine CRISPR's accuracy, which is essential work, given the ethical concerns.

THE MILLION GENOME PROJECT

the 30-second data

The Million Genome Project

(MGP), or All of Us, is the US National Institutes of Health initiative to unlock the genomes of 1 million Americans. The human genome has over 20,000 genes and is the DNA hereditary information that parents pass on to their child. MGP builds off the Human Genome Project (1990–2003), which created the world's first human DNA reference library, now used in medicine. Each person has a unique genome. Genes play a part in how we look (eye and hair colour) and act, as well as determine if we are predisposed to cancer or have genetic diseases. However, lifestyle and environment affect our health. MGP focuses on observing people's differences in health, lifestyle, environment and DNA. All of Us, uniquely, captures diversity – of people's backgrounds, environment from all regions across the country, and the broad spectrum of healthiness and sickness. Survey, electronic health record, physical measurement and biosample data will be collected to make one of the largest health databases for worldwide use. MGP will help develop more precise tools to identify, treat and prevent a person's disease by factoring how their health is affected by age, race, ethnicity, diet and environment – a concept known as precision medicine.

3-SECOND SAMPLE
The Million Genome Project will unlock the genome of 1 million Americans so that the data can be used to 'personalize' medicine.

3-MINUTE ANALYSIS
The Million Genome Project is a part of the US Government's Precision Medicine Initiative. Precision medicine, by using a person's genetic information, better tailors treatments, especially for diseases with a genetic component, such as Parkinson's disease. This healthcare approach enters into a new era of medicine where disease risk and treatment success can be more accurately predicted for each individual. Healthcare prevention can be better prioritized and cost-effective in this dawn of precision medicine.

RELATED TOPIC
See also
PERSONALIZED MEDICINE
page 138

3-SECOND BIOGRAPHIES
FRANCIS CRICK & JAMES WATSON
1916–2004 & 1928–
Co-discovered the structure of DNA in 1953 and won the Nobel Prize (1962).

FRANCIS COLLINS
1950–
Led the Human Genome Project (1990–2003) and has discovered genes associated with many different diseases.

30-SECOND TEXT
Rupa R. Patel

Advances in technology and data science have made the gathering and analysis of such large data sets possible.

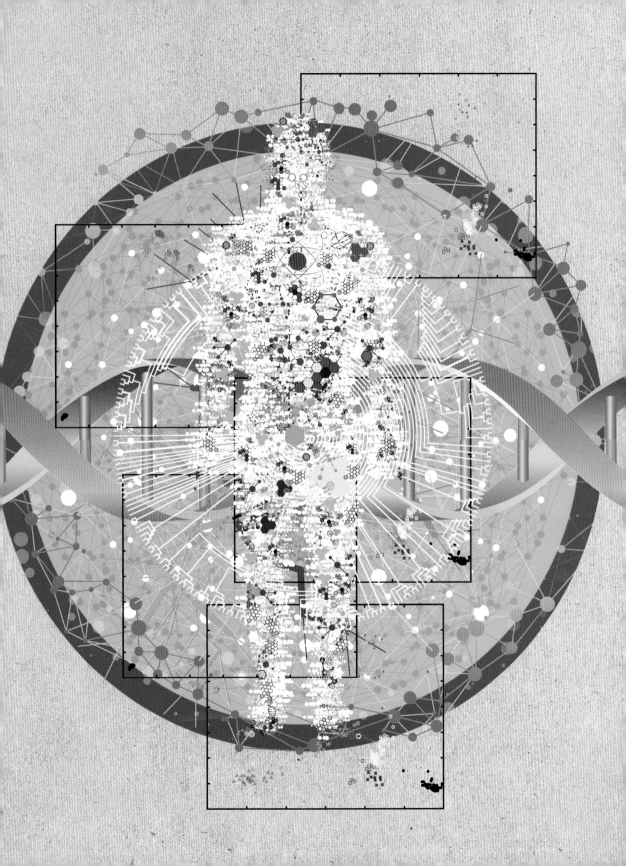

13 January 1900
Born in Dayton, Iowa, USA

1918
Graduates from high school

1929
Receives a Bachelor of Science degree from Iowa State College

1931
Receives the first Master of Science degree in Statistics from Iowa State College

1939
Appointed Assistant Professor of Statistics at Iowa State College

1940
Appointed Professor of Statistics at North Carolina State University (Raleigh)

1947
Founds Biometric Society

1949
Becomes first woman elected to the International Statistics Institute

1956
Elected as the President of the American Statistical Association

1959
Receives the O. Max Gardner Award from the University of North Carolina

1975
Invited to join the National Academy of Sciences

17 October 1978
Dies in Durham, North Carolina, USA

GERTRUDE COX

Methodism, crafts, psychology,
maths – this is a rare blend in today's zeitgeist
but one that Gertrude Cox pursued with zest
before and during her college years at Iowa
State. It is uncertain how this combination
of interests correlated with Cox's pursuit of
statistics thereafter, but her dissertation title
hints at its influence. After submitting her
thesis, 'A Statistical Investigation of a Teacher's
Ability as Indicated by the Success of His
Students in Subsequent Courses', Cox obtained
the first ever Master's degree awarded by Iowa
State. She then made her way to Berkeley to
pursue research in Psychological Statistics,
an endeavour that was abbreviated at the
insistence of her former mentor, George
Snedecor. Cox's former calculus professor as
well as employer, Snedecor sought her help in
organizing his statistical laboratory, and so Cox
traced her way back from Berkeley.

As a teacher, Cox had a knack for connecting
real-world research to course design. In 1939,
she was appointed Assistant Professor of
Statistics at Iowa State College and in 1940
began as the head of the new department
of Experimental Statistics in the School of
Agriculture and later ended up as the Director
of the Institute of Statistics at North Carolina
State. While there, Cox brought about several
innovations such as bringing in applied
statisticians to teach basic statistics ideas,
including a database of experimentation results
from different fields, and holding week-long
conferences on specific topics. Consequently,
she was the first woman elected to the
International Statistical Institute in 1949. As
an administrator, she was hugely successful
in securing grants to expand the scope of
offerings in her department. For example, in
1944 she received a grant from the General
Education Board to fund the establishment of
an Institute of Statistics. But Cox's repertoire
of roles did not end here. As an entrepreneur,
she pursued consulting assignments locally
and abroad; she also helped run a florist shop.
As a scholar and writer, she co-authored books
on experimental design; she also founded the
Biometric Society, serving as its first editor
and later as its president.

Cox is remembered for her contributions
to the fields of psychological statistics and
experimental design. In her biographical memoir
Richard Anderson, her friend and colleague,
says, 'Both as a teacher and a consultant,
Gertrude particularly emphasized randomization,
replication and experimental controls as
procedures essential to experimental design' –
tools that remain as vibrant as her batik pieces.

Aditya Ranganathan

CLIMATE CHANGE
the 30-second data

Climate trend predictions ensue

after compiling and processing volumes of data: average global temperatures over the years, for example. Average global temperature is a nuanced function of variables. Above-average build-ups of greenhouse gases in the atmosphere trap above-average amounts of heat, creating a barrier to prompt disposal. Other factors that slow down rates of heat emission include rising ocean levels, asphalt levels and decreasing ice. The result of this retardation is an upset of equilibrium – the desired state in which the rate of heat absorption equals the rate of heat emission, and average global temperature stays constant. Even though the disequilibrium is temporary, it is a period when heat lingers. And, when equilibrium returns, rather than catching up to the earlier temperature, we find ourselves in the midst of a new normal. There is a range of new 'normals' we could potentially reach: some mildly uncomfortable, some deadly. In order to understand which of these scenarios we might be heading towards, we must gather data vast enough to average out small fluctuations that might lead to incorrect predictions. The data that researchers are amassing includes global temperatures, sea-ice levels and so on – the conglomerate definitively indicating dangerous levels of greenhouse gas production.

3-SECOND SAMPLE
A firm prediction of the future of our planet rests upon the collection and analysis of massive amounts of data on global temperatures and greenhouse gas concentrations.

3-MINUTE ANALYSIS
Anthropogenic contributions, including expanding agricultural and industrial practices, correlate with an increase in global greenhouse gas concentrations and rising global temperatures, also known as global warming or climate change. The more data that is collected on anthropogenic contributions, the more conclusive the claim that it is human factors driving Earth's temperature change.

RELATED TOPIC
See also
CORRELATION
page 42

3-SECOND BIOGRAPHIES
JAMES HANSEN
1941–
NASA scientist and climate change advocate.

RICHARD A. MULLER
1944–
Climate sceptic converted to climate change advocate.

AL GORE
1948–
Published what was at the time a controversial documentary on the impacts of climate change called *An Inconvenient Truth*.

30-SECOND TEXT
Aditya Ranganathan

So far, the data collected has led 98 per cent of scientists to conclude that anthropogenic factors are to blame for climate change.

CURING CANCER

the 30-second data

While discoveries in basic science help explain the mechanisms of cancer, it is how these discoveries lead to targeted therapies and studies on patient outcomes that provide a deeper understanding of successful therapies and gets us closer to a cure. Data science allows us to test the value of intervention. Specifically, statistical thinking played a fundamental role in randomized trials, used for the first time by the US National Cancer Institute in 1954 to test treatments of patients with acute leukaemia. As long as 40 years ago, cancer research depended on many of the tasks that today define data science: study design, data analysis and database management. Today, molecular biology technologies produce thousands of measurements per patient, which permit the detection of mutations, structural chromosomal changes, aberrant gene expression, epigenetic changes and immune response in cancer cells. A primary aim is finding ways of using this information to improve diagnosis and develop tailored treatments. These new technologies produce large and complex data sets that require sophisticated statistical knowledge as well as computing skills to effectively work with the data and avoid being fooled by patterns arising by chance.

RELATED TOPIC
See also
HEALTH
page 92

3-SECOND SAMPLE
Advances in data science will be instrumental to curing cancer: they help us to understand if and why cancer interventions are working.

3-MINUTE ANALYSIS
Many challenges need to be overcome to cure cancer, and data will play a role in all of these. For example, it can take 10–15 years for a new drug to go through clinical trials and cost in excess of £1 billion. Using data science to optimize these processes for both money and time, while keeping them safe, is not usually how data and curing cancer are synthesized, but has become an important aspect of this work.

3-SECOND BIOGRAPHY
MARVIN ZELEN
1927–2014
Founding chair of what today is the Department of Data Sciences in the Dana-Farber Cancer Institute who developed many of the statistical methods and data management approaches used in modern clinical cancer trials.

30-SECOND TEXT
Rafael Irizarry

Data science has become crucial in cancer research and will play a key role in future advances.

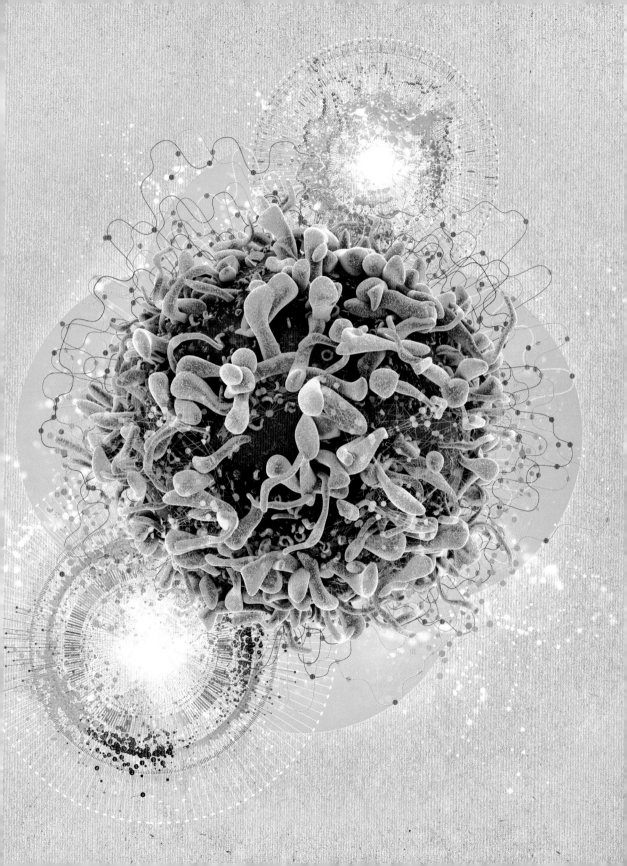

EPIDEMIOLOGY

the 30-second data

3-SECOND SAMPLE
There is an Ebola outbreak in Africa; what happens next? Epidemiology is used to collect data and study the who, what, where, when and why of the disease.

3-MINUTE ANALYSIS
Epidemiologic research is used to improve health by examining the causal relationship between risk factors (e.g. age, smoking) and disease (e.g. cancer, diabetes). Methods use observations or experiments, combined with statistics, to identify bias and false cause–effect associations. Milestones in health prevention occurred in the 1950s, when large epidemiologic studies provided conclusive evidence that tobacco smoking increased the risk of death from lung cancer and heart attacks.

Epidemiology is the science of collecting data and calculating how diseases are distributed, have patterns and are caused among people. The science blends multiple disciplines (i.e. statistics, social sciences, biology and engineering) together to create these calculations. The calculations are used to prevent and control both contagious and non-contagious diseases within populations. Epidemiology impacts public health and generates the evidence for the preventive (e.g. vaccines) and non-preventive procedures (e.g. diabetes screening) used today and will adopt tomorrow, such as microbiome-based diagnostics. Epidemiological evidence drives the health policies and guidelines that governments put in place, such as child vaccinations, to protect its citizens' health. The field is known for solving epidemics, or outbreaks of infectious diseases. Dr John Snow first defined epidemiologic concepts when he traced a contaminated water source to a cluster of cholera cases in London in 1854. Similarly, a group of deaths in Western Africa in 2013 led to an investigation to determine how and why the Ebola virus was spreading so quickly. The investigation informed health prevention programmes in the region to contain the virus's spread.

RELATED TOPICS
See also
STATISTICS & MODELLING
page 30

CORRELATION
page 42

HEALTH
page 92

3-SECOND BIOGRAPHIES
HIPPOCRATES
c. 460–370 BCE
First person to use the term 'epidemic' and observe how disease spread.

JOHN SNOW
1813–58
Successfully traced the source of the cholera outbreak in London in 1854, which went on to change urban water and sewage systems and public health worldwide; considered the father of epidemiology.

30-SECOND TEXT
Rupa R. Patel

Epidemiology enables calculations that are essential to our individual and collective well-being.

SOCIETY

aggregate statistics Statistics calculated across a set or group of data points, e.g. weekly sales by item.

anonymization Removing any information from a data set that could be used to identify or locate individuals, including names and addresses. True anonymization is difficult to achieve, as many variables such as location may allow individuals to be identified.

AI (artificial intelligence) Often used interchangeably with 'machine learning'. The process of programming a computer to find patterns or anomalies in large data sets, or to find the mathematical relationship between some input variables and an output. AI algorithms have applications in a range of fields including healthcare, self-driving cars and image recognition.

biometric airport security The use of biometric information, such as facial measurements or fingerprints, in airport security.

Brexit The exit of the United Kingdom from the European Union.

census A regular, systematic survey of members of a population, usually conducted by a government. Data collected during a census may include household size and income, and may be used to plan housing, healthcare and social services.

continuous health data Collected at regular, short intervals from individuals, and could include heart rate, activity or blood pressure. Advances in wearable technologies such as activity monitors make continuous health monitoring feasible.

differential privacy Method for sharing summary statistics about a group of people, while protecting the anonymity of individuals in the group.

geospatial data Involves a geographic component, which could include latitude and longitude or a country code.

Go Two-player strategy game, where the aim is to capture the most territory. Google's DeepMind has developed several algorithms designed to compete against humans.

Jeopardy! Televised American game show. Contestants are given answers, and must provide the correct questions.

machine learning Finding a mathematical relationship between input variables and an output. This 'learned' relationship can then be used to output predictions, forecasts or classifications given an input. For example, a machine learning model may be used to predict a patient's risk of developing diabetes given their weight. This would be done by fitting a function to a 'training' set of thousands of historic data points, where each point represents a single patient's weight and whether they developed diabetes. When a new, previously unseen patient's weight is run through the model, this 'learned' function will be used to predict whether they will develop diabetes. Modern computer hardware has enabled the development of powerful machine learning algorithms.

microtargeting Strategy used during political or advertising campaigns in which personalized messaging is delivered to different subsets of customers or voters based on information that has been mined or collected about their views, preferences or behaviours.

profile (voter) Information about an individual voter which may include age, address and party affiliation.

randomized experiments Experimental design where participants or subjects are randomly allocated to treatment groups. Participants in a randomized drug trial could be randomly allocated to a group where they would either receive a placebo or a drug.

sensitive information/data Reveals personal details, such as ethnicity, religious and political beliefs, sexual orientation, trade union membership or health-related data.

sniffers Software that intercepts and analyses the data being sent across a network, to or from a phone, computer or other electronic device.

Yellow Vests movement Protest movement originating in France, focused on issues such as rising fuel prices and the cost of living.

SURVEILLANCE

the 30-second data

Data surveillance is all around us, and it continues to grow more sophisticated and all-encompassing. From biometric airport security to grocery shopping, online activity and smartphone usage, we are constantly being surveilled, with our actions and choices being documented into spreadsheets. Geospatial surveillance data allows marketers to send you tailored ads based upon your physical, real-time location. Not only that, it can also use your past location behaviour to predict precisely what kind of ads to send you, sometimes without your permission or knowledge. While data surveillance is itself uninteresting; it's the actions taken from analysis of the data that can be both harmful and helpful. Using data surveillance, private and public entities are investigating methods of influencing or 'nudging' individuals to do the 'right' thing, and penalizing us for the doing the 'wrong' thing. A health insurance company could raise or lower rates based upon the daily steps a fitness tracker records; a car insurance company could do the same based upon data from a smart car. Data surveillance is not only about the present and analysis of actions; it's also about predicting future action. Who will be a criminal, who will be a terrorist, or simply, what time of the day are you most likely to buy that pair of shoes you have been eyeing while online shopping?

3-SECOND SAMPLE
Eyewitness sketches and background checks might become an archaic relic with the amount of surveillance data we now have the capability of storing and analysing.

3-MINUTE ANALYSIS
While data surveillance can feel negative, there are incredible advances in preventing terrorism, cracking child pornography rings by following images being sourced from the internet, and even aiding the global refugee crisis. The Hive (a data initiative for USA for the UN Refugee Agency) used high-resolution satellite imagery to create a machine-learning algorithm for detecting tents in refugee camps – allowing for better camp planning and field operation.

RELATED TOPICS
See also
SECURITY
page 84

MARKETING
page 108

3-SECOND BIOGRAPHY
TIM BERNERS-LEE
1955–
Creator of the World Wide Web, coining the internet as the 'world's largest surveillance network'.

30-SECOND TEXT
Liberty Vittert

When put towards a good cause, such as crime prevention, certain types of surveillance can be well justified.

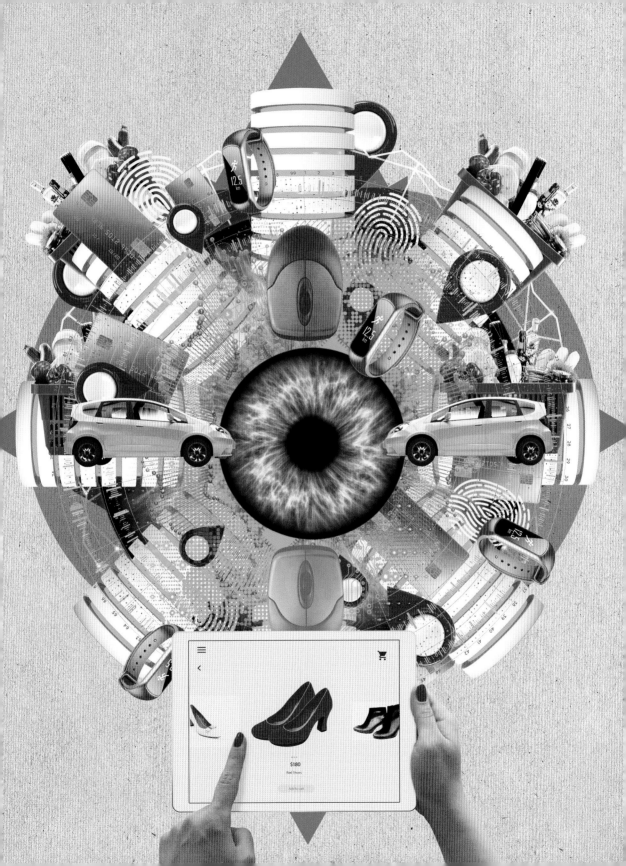

SECURITY

the 30-second data

Data is opening up new opportunities in intelligence processing, dissemination and analysis while improving investigative capacities of security and intelligence organizations at global and community levels. From anomalies (behaviour that doesn't fit a usual pattern) to association (relationships that the human eye couldn't detect) and links (social networks of connections, such as Al-Qaeda), intelligence organizations compile data from online activity, surveillance, social media and so on, to detect patterns, or lack thereof, in individual and group activity. Systems called 'sniffers' – designed to monitor a target user's internet traffic – have been transformed from simple surveillance systems to security systems designed to distinguish between communications that may be lawfully intercepted and those that may not for security purposes. Data can visualize how violence spreads like a virus among communities. The same data can also predict the most likely victims of violence and even, supposedly, the criminals. Police forces are using data to both target and forecast these individuals. For example, police in Chicago identified over 1,400 men to go on a 'heat list' generated by an algorithm that rank-orders potential victims and subjects with the greatest risk of violence.

RELATED TOPICS
See also
SURVEILLANCE
page 82

ETHICS
page 152

3-SECOND BIOGRAPHY
PATRICK W. KELLEY
fl. 1994
FBI Director of Integrity and Compliance, who migrated Carnivore to practice.

30-SECOND TEXT
Liberty Vittert

3-SECOND SAMPLE
Big Data meets Big Brother in the untapped and untried world of data-driven security opportunities. From community policing to preventing terrorism, the possibilities are endless, and untested.

3-MINUTE ANALYSIS
In the case of Chicago (see 'data' text), the higher the score means the greater risk of being a victim or perpetrator of violence. In 2016, on Mother's Day weekend, 80 per cent of the 51 people shot over two days had been correctly identified on the list. While proponents say that it allows police to prioritize youth violence by intervening in the lives of those most at risk, naysayers worry that by not identifying what generates the risk score, racial bias and unethical data use might be in practice.

Carnivore was one of the first systems implemented by the FBI to monitor email and communications from a security perspective.

PRIVACY

the 30-second data

3-SECOND SAMPLE
Every day we generate thousands of data points describing our lifestyle and behaviour. Who should have access to this information, and how can they use it responsibly?

3-MINUTE ANALYSIS
Governments have taken steps to safeguard privacy. The UK's Information Commissioner's Office fined Facebook £500,000 for failing to protect user data. In the European Union, organizations must ask for consent when collecting personal data and delete it when asked. The US Census Bureau is introducing 'differential privacy' into the 2020 census, a method that prevents individuals being identified from aggregate statistics.

The adage 'if you're not paying for the product, you are the product' remains true in the era of big data. Businesses and governments hold detailed information about our likes, health, finances and whereabouts, and can harness this to serve us personalized advertising. Controversies around targeted political campaigning on Facebook, including alleged data breaches during the 2016 US presidential election, have brought data privacy to the forefront of public debate. For example, medical records are held securely by healthcare providers, but health apps are not subject to the same privacy regulations as hospitals or doctors. A *British Medical Journal* study found that nearly four in five of these apps routinely share personal data with third parties. Users of menstrual cycle, fitness or mental health tracking apps may be unaware that sensitive information about their health and well-being is up for sale. One strategy for protecting privacy is the removal of identifying variables, such as full names or addresses, from large data sets. But can data ever truly be anonymized? In 2018 the *New York Times* reviewed a large anonymized phone location data set. Journalists were able to identify and contact two individuals from the data, demonstrating that true anonymization is difficult to achieve.

RELATED TOPICS
See also
SURVEILLANCE
page 82

REGULATION
page 150

3-SECOND BIOGRAPHY
MITCHELL BAKER
1959–
Founder of the Mozilla Foundation, launched in 2003, which works to protect individuals' privacy while keeping the internet open and accessible.

30-SECOND TEXT
Maryam Ahmed

Non-governmental organizations advocate for and support projects relating to greater internet and data privacy.

12 May 1820
Born in Italy, and is
named after the city
of her birth

1837
Experiences the first of
several 'calls from God',
which inform her desire
to serve others

1844
Announces intention
to pursue a career in
nursing, prompting
opposition from family

1851
Undertakes medical
training in Düsseldorf,
Germany

1853
Becomes Superintendent
of the Institute for
the Care of Sick
Gentlewomen, London

1854
Travels to the Scutari
Barracks in modern-day
Turkey, with a group of
38 female nurses, and
oversees the introduction
of sanitary reforms

1857
Suffers from intermittent
episodes of depression
and ill health, which
continue until her death

1858
Publishes the data-driven
report 'Mortality of the
British Army'

1859
Elected to the Royal
Statistical Society

13 August 1910
Dies in her sleep,
in London

FLORENCE NIGHTINGALE

Florence Nightingale was a pioneer of modern nursing methods, medical statistics and data visualization. Born to a wealthy British family in 1820, she defied the cultural conventions of the time by choosing to pursue her calling as a nurse rather than simply marrying and raising a family.

After training in Germany, Nightingale, aged 33, rose to become the superintendent of a hospital in London. It was during the Crimean War, however, that she would make her mark on nursing and data science. In 1854, at the invitation of the Minister for War, Florence travelled to a military hospital in Scutari, Turkey, to supervise the introduction of female nurses. Under Nightingale's watch, the death rate fell, partly due to her emphasis on basic sanitary practices such as handwashing. Here, her habit of performing ward rounds at night earned her the affectionate moniker 'the Lady with the Lamp'.

Throughout her career, Nightingale took a data-driven approach to her nursing practice. She made extensive use of data visualizations and statistics to highlight public health issues.

During her time in the Crimea, Nightingale collected data on mortality rates and causes of deaths. In 1858, she published 'Mortality of the British Army', a report demonstrating the differences between death rates in the military compared to the civilian population. She was an early adopter of the coxcomb, a form of pie chart where the length rather than the angle of each segment is proportional to the size of the data. In one of her best known visualizations, Nightingale used a coxcomb to illustrate that many more soldiers died of 'preventable or mitigable' diseases than of fatal wounds.

Nightingale was also concerned with the living conditions of British soldiers in India, and drove the establishment of a Royal Commission to investigate the issue. In 1873, she reported that the death rate for British soldiers in India had fallen from 69 to 18 per 1,000 following ten years of sanitary reform.

In Britain, Nightingale lobbied government ministers for reforms including compulsory sanitation in private houses, improved drainage and stricter legislation.

Nightingale's contributions to medical statistics were widely recognized. In 1859, she became the first female member of the Royal Statistical Society, and later became an honorary member of the American Statistical Association. She died in her sleep in 1910, aged 90.

Maryam Ahmed

VOTE SCIENCE

the 30-second data

Vote Science has been in practice since political outcomes began being decided by votes, dating back to sixth-century BCE Athens. Modern Vote Science evolved rapidly in the US in the 1950s, when campaigns, political parties and special interest groups started keeping large databases of eligible voters, which were later used to build individual voter profiles. Using machine learning and statistical analysis, campaign professionals began using these profiles to make calculated decisions on how to win an election or sway public opinion. Current best practices include maintaining databases of people with hundreds of attributes, from individuals' credit scores to whether they vote early/in-person or even if they are more likely to vote if reminded via phone, text or email. Using this data, campaigns and political parties work to predict voter behaviour, such as whether voters will turn out, when they will vote, how they will vote and – most recently – what will persuade them to change their opinion. Recent campaigns have adopted randomized field experiments to assess the effectiveness of mobilization and persuasion efforts. Vote Science now determines how a campaign chooses to spend its advertising funds as well as which particular messages are shown to specific, individual voters.

3-SECOND SAMPLE
Vote Science is the practice of using modern voter registration lists, consumer and social media data, and polling to influence public opinion and win elections.

3-MINUTE ANALYSIS
George Bush's 2004 re-election was the first political campaign to use political microtargeting – the use of machine-learning algorithms to classify voters on an individual level of how they might vote or if they even would vote. Barack Obama's campaigns in 2008 and 2012 took Vote Science a step further by incorporating randomized field experiments. Elections in the UK, France and India began to use Vote Science techniques such as microtargeting and random field experiments in their campaigns after witnessing the success of the American model.

RELATED TOPICS
See also
LEARNING FROM DATA
page 20

ETHICS
page 152

3-SECOND BIOGRAPHIES
DONALD P. GREEN
1961–
Leader of Vote Science randomized experiments.

SASHA ISSENBERG
fl. 2002–
Chronicler of how data science has been used and evolved in campaigns in the last 20 years.

DAN WAGNER
fl. 2005–
Director of Analytics for 'Obama for America' in 2012; led efforts to expand Vote Science in campaigns to message testing and donor models.

30-SECOND TEXT
Scott Tranter

Modern-day election campaigns are driven by Vote Science, with a vast amount of campaign budget allocated to it.

HEALTH
the 30-second data

Data science develops tools to analyse health information, to improve related services and outcomes. An estimated 30 per cent of the world's electronically stored data comes from the healthcare field. A single patient can generate roughly 80 megabytes of data annually (the equivalent of 260 books worth of data). This health data can come from a variety of sources, including genetic testing, surveys, wearable devices, social media, clinical trials, medical imaging, clinic and pharmacy information, administrative claim databases and national registries. A common data source is electronic medical record (EMR) platforms, which collect, organize and analyse patient data. EMRs enable doctors and healthcare networks to communicate and coordinate care, thereby reducing inefficiencies and costs. EMR data is used to create decision tools, for clinicians, which incorporate evidence-based recommendations for patient test results and prevention procedures. Healthcare data science combines the fields of predictive analytics, machine learning and information technology to transform unstructured information into knowledge used to change clinical and public health practice. Data science helps to save lives by predicting patient risk for diseases, personalizing patient treatments and enabling research to cure diseases.

RELATED TOPICS
See also
EPIDEMIOLOGY
page 76

PERSONALIZED MEDICINE
page 138

MENTAL HEALTH
page 140

3-SECOND BIOGRAPHIES
FLORENCE NIGHTINGALE
1820–1910
Championed the use of healthcare statistics.

BILL & MELINDA GATES
1955– & 1964–
Launched in 2000, the Gates Foundation uses data to solve some of the world's biggest health data science problems.

JAMES PARK & ERIC FRIEDMAN
fl. 2007
Founders of Fitbit who applied sensors and wireless tech to health and fitness.

30-SECOND TEXT
Rupa R. Patel

3-SECOND SAMPLE
Data science transforms unstructured health information into knowledge that changes medical practice.

3-MINUTE ANALYSIS
Consumer-grade wearable devices coupled with smartphone technology offer innovative ways to capture continuous health data, improving patient outcomes. For example, heart monitors can be used to diagnose and/or predict abnormal and potentially life-threatening heart rhythms. The data can be assessed by varying time parameters (days to weeks versus months to years), to develop early-warning health scores. Similarly, hearing aids with motion sensors can detect the cause of a fall (slipping versus heart attack), so doctors can respond effectively.

Using data to personalize healthcare helps to save lives.

IBM'S WATSON & GOOGLE'S DEEPMIND

the 30-second data

3-SECOND SAMPLE
IBM's Watson *Jeopardy!*-playing computer and Google's DeepMind Go-playing program introduced the world to machine learning and artificial intelligence in ways that were easy to understand.

3-MINUTE ANALYSIS
Computer companies pursue targets such as playing *Jeopardy!* and Go because to excel at them they have to develop general-purpose capabilities that can be applied to other commercially important problems. The ability to answer a person's question in his or her own language on a broad range of topics, or to train for complicated problems such as robot navigation, will help future computers to perform more sophisticated tasks for people, including their creators.

When IBM's Watson computer defeated the reigning *Jeopardy!* champion on a nationally televised game show in 2011, it was a demonstration of how computer-based natural language processing and machine learning had advanced sufficiently to take on the complex wordplay, puns and ambiguity that many viewers might struggle with. Google's DeepMind subsidiary did something similar – its AlphaGo program used machine learning and artificial intelligence to beat the world champion at Go, a very complicated strategy board game played with black and white stones; a feat no other computer had ever accomplished. Picking ambitious targets such as beating humans at well-known games serves several purposes. First, it gives data scientists clear goals and benchmarks to target, like 'Win at *Jeopardy!*'. In IBM's case, they even announced the goal beforehand, which put pressure on the development team to be creative and think outside the box, as who would want to be publicly humiliated by a mere human? Second, these sparring matches speak to the public about how far hardware and software are progressing. Go is much more challenging than chess, so if a computer can beat the world champion, we must be making a lot of progress!

RELATED TOPICS
See also
MACHINE LEARNING
page 32

NEURAL NETWORKS & DEEP LEARNING
page 34

GAMING
page 130

3-SECOND BIOGRAPHIES
THOMAS WATSON
1874–1956
Chairman and CEO of IBM, the *Jeopardy!*-playing computer is named after him.

DEEPMIND TECHNOLOGIES
2010–
Acquired by Alphabet (parent of Google) in 2014.

30-SECOND TEXT
Willy Shih

Computers beating humans at ever more complex games is a visible measure of the progress being made in data science.

BUSINESS ◑

automated system Some repetitive tasks or calculations can be carried out faster, continuously and more efficiently by computers. Examples of automated systems include automated passport gates at airports, self-driving cars or speech-to-text software.

autonomous machines Able to complete a task without human input, such as a self-driving car.

big data Data set that meets some or all of the following criteria; volume, velocity, veracity and variety; must consist of a large volume or amount of individual data points, generated at a high or regular velocity. It may consist of a variety of data types including text, numerical data or images, and it will ideally be accurate or have veracity.

data analytics Obtaining, cleaning and analysing data to gain useful insights, answer research questions or inform decision making. Prescriptive data analytics describes and draws conclusions from the available data; predictive analytics aims to generalize these findings to make predictions or forecasts about the future.

foot traffic analysis Often used in the retail sector to measure how many customers enter a shop, and their movements and behaviour while browsing.

geolocation data Describes the location of a person or object over time.

Go Two-player strategy game, where the aim is to capture the most territory. Google's DeepMind has developed several algorithms designed to compete against humans.

Internet of things (IoT) Internet-connected or 'smart' devices, including activity monitors, home assistants and TVs, which provide improved functionality compared to their offline counterparts through the collection and analysis of data in real time. For example, a smart home assistant is able to communicate with and control items around the home such as smart light bulbs, central heating and security systems.

natural language-processing algorithms
Techniques for analysing written or spoken language. This could include the contents of political speeches, vocal commands given to a smartphone or written customer feedback on an e-commerce website. Common natural language processing techniques include sentiment analysis, where text is labelled as positive or negative depending on its tone, and topic modelling, which aims to identify the overall theme or topic of a piece of text.

probability theory Branch of mathematics concerned with representing probabilities in mathematical terms. The field relies on a set of underlying assumptions, or axioms, including 'the probability of an event is a non-negative, real number.'

prototype Working draft version of a piece of software or hardware, sometimes referred to as a minimum viable product, or MVP.

quantitative finance Uses probability and mathematical techniques to model financial problems.

quantum mechanics Branch of physics concerned with the behaviour of atomic and subatomic particles.

reinforcement learning Branch of machine learning, where algorithms learn to take actions which maximize a specified reward.

tabulating system A machine, developed in the 1800s, designed to store information in the form of hole-punched cards. Its first use was in the 1890s, to store data collected during the first US census.

tracking cookies Piece of information from a website, stored by a person's web browser, which is shared or tracked across websites, to track a user's online journey. They may be used by third party advertising providers, to serve personalized adverts based on a user's browsing history.

INDUSTRY 4.0

the 30-second data

Industry 4.0 can be more easily understood as a 'smart factory', where internet-connected systems/machines communicate and cooperate with each other in real-time to do the jobs that humans used to do. This relies on the Internet of things (IoT), the extension of internet connectivity into devices and everyday objects. While Industry 4.0 can have an ominous ring to it in certain circles, there is a vast amount of incredible applications to our daily lives. From robots picking and packing items in a warehouse for delivery, to autonomous cranes and trucks on building sites, and using information collected from these machines to find and optimize irregularities in business systems – the possibilities are endless and, as of yet, unknown. Business is not the only winner in this industrial revolution. For example, providing assistance to elderly or disabled individuals through homecare advances with systems like voice control or alerts for falls or seizures. However, there are large barriers to the full implementation of Industry 4.0, integration being one of the biggest. There are no industry standards for connectivity and the systems themselves are fragmented between different industries and companies. Privacy concerns are overwhelming, with the amount of data collected (personal and otherwise) by these systems needing to be protected, as are decisions over ownership.

3-SECOND SAMPLE
'Humankind will be extinct or jobless' is the feared mantra with the fourth industrial revolution in manufacturing, where machines use data to make their own decisions.

3-MINUTE ANALYSIS
Millions of people are employed by the manufacturing industry and fears over job loss from the data revolution/ Industry 4.0 are real and already evident. While this may be very worrisome, many are using it as an opportunity to push for the idea of a 'universal basic income'. This is a periodic monetary compensation given to all citizens as a right, with the only requirement being legal residency. This stipend would be enough for basic bills and living, with the aim that individuals will be free to pursue any interest.

RELATED TOPIC
See also
ARTIFICIAL INTELLIGENCE (AI)
page 148

3-SECOND BIOGRAPHY
HUGH EVERETT
1930–82
First proposed the Many Worlds interpretation for quantum mechanics and operations research.

30-SECOND TEXT
Liberty Vittert

Connectivity and standardization across industries are a major obstacle to the widespread adoption of smart factories.

ENERGY SUPPLY & DISTRIBUTION

the 30-second data

3-SECOND SAMPLE
Data science is key to managing the growth of renewable and distributed energy sources in the electric power system.

3-MINUTE ANALYSIS
Fossil fuels still make up a large part of global energy consumption, and oil and gas companies make liberal use of analytics as well – in characterizing untapped oil reservoirs below the earth's surface, optimizing drill operations when drilling new wells, forecasting impending equipment failures, deciding which oil streams to blend together, and more.

Our energy supply is transitioning from fossil fuels and centralized infrastructure to a renewable, decentralized system, and data analytics eases the challenges of that transition. As the output of wind farms and solar photovoltaic plants is weather-dependent, high-resolution weather forecasting based on predictive analytics has wide applications in improving design and operation of these systems, from optimizing the layout of wind turbines in a field to automatically adjusting the angle of solar panels to maximize power generation despite changing conditions. As electricity is then transmitted to the end customer, analytics is critical to managing the growing complexity of the power grid due to 'distributed energy resources' – controllable devices such as backup generators, home batteries and smart thermostats, often owned by homeowners and businesses. These devices are excellent resources for balancing the grid, and grid operators can use analytics to determine which mix of devices to pull from at any time based on weather, historic energy demand, the performance and tolerances of each device, and grid conditions like voltage. For grid operators, analytics is also useful in planning infrastructure investments, allowing them to predict which parts of the network will be most strained decades into the future.

RELATED TOPICS
See also
INDUSTRY 4.0
page 100

3-SECOND BIOGRAPHY
THOMAS EDISON
1847–1931
Architect of the world's first power grid, which went live in New York City in 1882.

30-SECOND TEXT
Katrina Westerhof

Data like demographics and infrastructure condition can inform decisions about where to increase the capacity of the power grid.

LOGISTICS
the 30-second data

Route optimization – born of

both predictive and prescriptive data analytics – has unlocked enormous benefits for the previously low-tech logistics industry, reducing fuel consumption and improving reliability of service. When delivering packages to homes and businesses, logistics companies can now identify the most efficient routes for each driver, each day, across the entire fleet, taking into account delivery deadlines, traffic patterns and weather forecasts. Upstream, in freight shipping, shippers can apply similar techniques to optimize the route from an origin point to a distribution facility, choosing the right combination of sea, air, rail and road transport to get each shipment to its destination most efficiently and on time. In both cases, the tools exist today to make these optimizations dynamic, allowing carriers to reroute parcels in real time as conditions change, and for delivery routes, even recommending the ideal driving speed on each leg of the route to consistently hit green traffic lights. Beyond optimizing how an item gets to its destination, big data and analytics also provide insights into how to structure a global logistics network, such as where to build new hubs, distribution facilities and customer drop sites as transportation constraints and customer demand change.

RELATED TOPICS

See also
INDUSTRY 4.0
page 100

SHOPPING
page 118

3-SECOND BIOGRAPHY
JUAN PEREZ
1967–
Chief Engineering and Information Officer at UPS who led the implementation of the company's ORION route optimization project.

30-SECOND TEXT
Katrina Westerhof

3-SECOND SAMPLE
Getting an item from Point A to Point B is more efficient and reliable with optimized routing, enabled by data analytics.

3-MINUTE ANALYSIS
In the context of supply-chain management, the value of analytics for logistics is even greater. Predictive analytics will improve inventory management by considering the impacts of factors like geopolitics, weather and climate change, and consumer sentiment on product availability or demand. And integrating data across the supply chain unlocks new opportunities – for example, dynamically rerouting a shipment of ripe fruit to a nearer store or a store where fruit sells more quickly, thereby reducing food waste.

Dynamic route optimization enables shippers to be responsive to changing conditions in the supply chain.

29 February 1860
Born in Buffalo,
New York, USA

1875
Enrols in City College
of New York

1879
Receives undergraduate
degree in Mining from
Columbia University

1880
Serves as assistant to
William Trowbridge, his
professor who worked
on the US Census

1889
Receives patent for
punch-card tabulator
(Patent No. 395,782)

1890
Gains PhD from Columbia
University

1890
Receives the Elliot
Cresson Medal

1890–1900
Contracted to supply his
machines for the 1890
census count

1911
Begins (alongside several
others) the Computing-
Tabulating-Recording
Company (CRT)

1918
Starts stepping back from
day-to-day operations
at CRT

1921
Retires

1924
CRT becomes IBM

17 November 1929
Dies in Washington, DC,
USA

HERMAN HOLLERITH

On 2nd January, 1889, the US Patent Office awarded Patent No. 395,782 for an invention 'which consists in recording separate statistical items pertaining to the individual by holes or combinations of holes punched in sheets…and then counting [said items]…by means of mechanical counters'. Little did the Patent Office suspect that No. 395,782 would revolutionize the tracking of global populations for the next two decades and inspire the very creation of modern computation. But who was the author of that patent? Enter Herman Hollerith – brilliant statistician, consultant to the US census office and lover of chicken salad.

Born in 1860, Hollerith was a precocious child, though his hatred for spelling often cast a cloud over his early education. After earning an undergraduate degree at the University of Columbia, he was invited to work with professor W.P. Trowbridge in 1880. Over the next ten years, Hollerith dabbled in technology, taught at Massachusetts Institute of Technology (MIT) and served a brief stint in a patent office. But there was more to come.

As the story goes, one evening in 1882, Hollerith was having dinner with his girlfriend and her father, Dr John Shaw Billings, when Billings – who was then employed by the US Census Bureau, proposed the idea of an automated system for counting census votes. Hollerith's inventive fire was stoked, and building such a system became his preoccupation at MIT. Eventually, a tabulating system emerged, one which used a series of insulating punch cards, such that each hole corresponded to a particular census category.

Following a patent award in 1889, the US census adopted Hollerith's machines in the 1890 census, saving years and millions of dollars. Other countries, including Canada, Norway, Austria and England, were quick to adopt Hollerith's system. And the old ways, like the chad that falls out after punching, fell to the wayside.

Hollerith was not only a great statistician and inventor but also an entrepreneur. In 1896, he founded his Tabulating Machine Company to sell his machines. As competition emerged, Hollerith's company merged with other companies to form the Computer Tabulating Recording Company, which he served until 1911 in an advisory capacity. In 1924, the Computer Tabulating Recording Company became IBM, one of the largest computer companies in the world today.

From his inventions that aided governments and set American computing on a roaring path, to his statistical insights and mechanical know-how that solved large-scale data storage and management problems, Herman Hollerith has left his imprint.

Aditya Ranganathan

MARKETING
the 30-second data

The advent of marketing data science has allowed businesses to target their ideal customers more effectively. To some extent, this has had equalizing effects for up-and-coming retailers competing with large, established online retail platforms like Amazon. However, the established players still have an incredible competitive advantage in the form of user data. While smaller firms rely almost exclusively on tracking cookies and data brokers (who provide demographic and commercial information on individuals) to target a customer profile, large industry players such as Amazon, Google and Alibaba have immense amounts of data on every single one of their respective users. In a 2019 disclosure by Amazon, it was revealed that over 100 million customers pay a membership fee for Amazon Prime, its two-day shipping service bundled with myriad other services like streaming content and food delivery. The marketing advantage due to data availability alone is so pronounced in this case that antitrust legislation is being proposed to prevent companies from acting as both sellers and platforms for selling. However the future of online commerce plays out, it is clear that as society becomes ever more digital, data science will be an essential method in determining who we are and, more importantly, what we want to buy.

RELATED TOPICS
See also
PRIVACY
page 86

SOCIAL MEDIA
page 128

3-SECOND BIOGRAPHIES
SUSAN WOJCICKI
1968–
CEO of YouTube following Google's acquisition, and dubbed the 'most important person in advertising'.

JEFF HAMMERBACHER
1982–
Chief scientist and co-founder at Cloudera; formerly led the data science team at Facebook.

30-SECOND TEXT
Scott Tranter

3-SECOND SAMPLE
As data science rose from its infancy to become its own field, separate in discipline from software engineering or statistics, digital marketing's dominance rose with it.

3-MINUTE ANALYSIS
Some lament data science and marketing's close relationships. Cloudera's own Jeff Hammerbacher (formerly lead data scientist at Facebook) has been famously quoted: 'The best minds of my generation are thinking about how to make people click ads.' In its most nefarious form, data science is a key contributor to the social media addiction phenomenon, as practitioners work to keep you engaged with the platform by which they can serve you adverts for longer periods of time.

Modern marketing methods are driven by vast amounts of data, used to create highly targetable consumer profiles.

FINANCIAL MODELLING

the 30-second data

Ever since Ed Thorp popularized

the application of probability theory in financial markets, trying to beat the market has been a continuous, elusive conquest, and data science has been an integral tool in optimizing investment strategies. In the past few years, competition has been fierce for new alternative data sets, from which unique and proprietary insights can be extracted, hoping to give traders the edge in predicting the next price move. For example, aggregated credit card data can be used to estimate real-time revenues for companies ahead of earnings, which are otherwise released only quarterly, and bets can be placed using this granular source of information ahead of the official release. Foot traffic analysis can also help with estimating and trading around earnings – satellite imagery to count cars at supermarket car parks can indicate trends in volume of shoppers, and geolocation data derived from devices such as mobile phones can illuminate consumer behaviour. Natural language-processing algorithms enable machines to understand text, and can further help extract sentiment about the market when applied to news feeds, call transcripts regarding company earnings and analyst reports.

3-SECOND SAMPLE
With the quantitative revolution in finance, data scientists are en vogue, and potentially the key to the holy grail of figuring out the market's next move.

3-MINUTE ANALYSIS
Market impact takes place when a participant seeks to buy or sell an asset on an exchange, and their own footprint pushes the price adversely against them. For example, when buying 10,000 shares of Apple, the price of the last share purchased may be higher than the price when the trade was initiated. Minimizing this effect can be modelled as a classic 'reinforcement learning' problem, where feedback from a series of actions are incorporated into subsequent trading decisions.

RELATED TOPICS
See also
MACHINE LEARNING
page 32

IBM'S WATSON &
GOOGLE'S DEEPMIND
page 94

3-SECOND BIOGRAPHY
EDWARD O. THORP
1932–
US mathematics professor who pioneered the modern applications of probability theory in the financial markets.

30-SECOND TEXT
Sivan Gamliel

As availability of diverse data sources continues to grow, so grows the dominance of quantitative traders on Wall Street.

NEW PRODUCT DEVELOPMENT

the 30-second data

Developing a product means

solving a problem or fulfilling a desire that a customer is willing to pay for. Often, developers start by looking at products that sold in the past, then, for inspiration and prototyping, the product is tested with surveys or focus groups, after which it is tweaked accordingly and put on the market. Collecting data helps product developers to refine features and pricing, but usually intuition plays a big role. The internet has turned this process on its head, because every move a customer makes while browsing products online, reading the news or even watching TV can be tracked. Netflix collects a consumer's preferences on joining, along with ratings on films you have already seen. When using the service, it tracks when and what is watched, whether viewers return after a pause, whether they finish watching a programme and what they watch next. Before a new programme is developed, the developers already know – from data analytics on their millions of members – what each viewer likes and how many are likely to watch it. They then make programmes with actors and stories they know will be popular. This is just one example of how data analytics is changing product development.

RELATED TOPICS

See also
LEARNING FROM DATA
page 20

MARKETING
page 108

SHOPPING
page 118

30-SECOND TEXT
Willy Shih

3-SECOND SAMPLE
Internet-based data collection and analytics offers much more detailed data on when, what and how people use products, enabling a much deeper understanding of consumer needs.

3-MINUTE ANALYSIS
Big consumer products companies like Procter & Gamble use data to build computer models of new products such as disposable nappies even before their first physical prototype. When they go to market, they use a computerized 'digital pulse' to track blogs, tweets and online ratings and comments to see how their products are faring. This allows them to react quickly to things that happen in the marketplace, both good and bad.

Owing to the vast amount of information developers have at their fingertips, they might know what you want even before you do.

PLEASURE

PLEASURE
GLOSSARY

AI (artificial intelligence) Often used interchangeably with 'machine learning'. The process of programming a computer to find patterns or anomalies in large data sets, or to find the mathematical relationship between some input variables and an output. AI algorithms have applications in a range of fields including healthcare, self-driving cars and image recognition.

algorithm Set of instructions or calculations designed for a computer to follow. Writing algorithms is called 'coding' or 'computer programming'. The result of an algorithm could be anything from the sum of two numbers to the movement of a self-driving car.

analytical engine Mechanical computer, designed by Charles Babbage in the early 1800s, intended to carry out arithmetic and logical operations, taking instructions or inputs via hole-punched cards. The machine was not constructed during Babbage's lifetime, but a modified version was built by the London Science Museum in 1991.

cookies Pieces of information from a website, stored by a person's web browser, which may help the website remember information specific to that person, such as items added to an online shopping cart or login details.

correlated risk Multiple negative outcomes or losses, caused by a single event. For example, many homes are likely to be damaged and people injured as the result of a single hurricane.

data analytics Obtaining, cleaning and analysing data to gain useful insights, answer research questions or inform decision making.

digital age Time period beginning in the 1970s and stretching to the present day, characterized by rapid technological advances, including the introduction of the personal computer and the rise of the internet.

digital library Large repository or archive of data, sometimes available to access or download through the internet, for commercial or research purposes. Digital libraries may include images, text or numerical data.

esports Electronic sports in which individuals or teams of players compete in international tournaments and for monetary prizes, to win video games.

geolocated franchising model Teams of competitive video-game players, based in a specific city, can form a franchise to compete in international or national esports tournaments for a particular game.

live streaming The live broadcast of video or audio content, via the internet. Esports are usually watched through live streaming.

machine learning Finding a mathematical relationship between input variables and an output. This 'learned' relationship can then be used to output predictions, forecasts or classifications given an input.

metrics Quantitative measure of performance. For example, it is important to assess accuracy metrics for automated decision-making algorithms. Similarly, measures such as inflation or the FTSE 100 index could be seen as a performance metrics for the economy.

model/modelling Real world processes or problems in mathematical terms; can be simple or very complex, and are often used to make predictions or forecasts.

STEM The fields of science, technology, engineering and mathematics.

swipe The act of swiping a finger across a smartphone screen, to interact with an app. Swiping is widely used in dating apps, where users often swipe right or left on a photograph of a potential romantic partner, to signal interest or disinterest.

wearable technology Electronic devices that can be worn on the body including activity monitors and smart watches.

SHOPPING

the 30-second data

RELATED TOPICS
See also
DATA COLLECTION
page 16

LEARNING FROM DATA
page 20

ARTIFICIAL INTELLIGENCE (AI)
page 148

3-SECOND SAMPLE
Shopping online has changed shopping as we know it, but how is it possible that websites seem to know what we want before we do?

3-MINUTE ANALYSIS
Ever wondered how a website knows the shoes you were looking at the other day? Well, the answer is cookies. These are small pieces of data that come from a website and are stored in the web browser, allowing websites to remember various nuggets of information including past activity or items in a shopping cart, which explains why that pair of shoes just keeps coming back.

With the internet giving a home to a variety of retailers, the consumer can now buy almost anything from the comfort of their own home. The consequence of this is that retailers have been able to harvest extensive and accurate data relating to customers, which means they are better able to target shoppers based on their habits. An example of this can be seen on Amazon – the biggest online retailer in the world – with its ability to recommend items based on your previous purchases, ratings and wish lists. However, the ability to perform this type of activity is not only the realm of companies the size of Amazon. Services now exist offering artificial intelligence (AI) solutions that allow retailers of many sizes to be able to harness the power of these types of algorithms to drive business, which means that the next time an online retailer suggests a T-shirt to go with your jeans, it could be via AI. Data science isn't restricted to shopping suggestions: it also applies to how goods are purchased. Facial recognition technology combined with smart devices allows payments to be authenticated without the use of credit cards.

3-SECOND BIOGRAPHY
JEFF BEZOS
1964–
Tech entrepreneur who is the founder, CEO and president of Amazon.

30-SECOND TEXT
Robert Mastrodomenico

Online shopping and new payment methods combine to create a high-tech consumerism that's feeding valuable data to retailers.

DATING

the 30-second data

Sign up to a dating site and you're presented with a number of questions that you have to complete which will define you and find you your perfect match – how is this possible? The questions are weighted based on their importance, and using these as an input to the algorithms used allows a score to be calculated that shows your satisfaction with other potential matches. It's not all about you, though – the best match also takes into account how well your answers mesh with the potential matches. So the stats behind your match doesn't assume love is a one-way street, which seems sensible. Online dating also includes the generation of 'swipers' who use dating apps. Here, you are able to see potential matches based on fixed data such as location, age preference and so on. The application then shows you individuals and you register your thoughts on the individuals by swiping left or right. Who appears on your screen is not just based on your fixed preferences; instead, complex algorithms learn from how you and others have used the app and sends you individuals who you would most likely respond to positively.

RELATED TOPICS
See also
REGRESSION
page 24

CLUSTERING
page 28

MACHINE LEARNING
page 32

3-SECOND SAMPLE
Online dating has changed the game of finding love to the extent that finding 'the one' is more statistical than you may think.

3-MINUTE ANALYSIS
The 'swipe right' paradox: should dating app users just keep swiping to see everyone on the application? Given that the best selections will come first, every subsequent swipe should give a worse selection, and eventually you will see recycled selections as someone you said no to is better than a very bad match, at least from a mathematical point of view.

3-SECOND BIOGRAPHIES
DAVID SPIEGELHALTER
1953–
Statistician, Professor of the Public Understanding of Risk and author of *Sex by Numbers*.

HANNAH FRY
1984–
Mathematician, lecturer, writer and TV presenter who studies patterns of human behaviour in relation to dating and relationships.

30-SECOND TEXT
Robert Mastrodomenico

Love at first swipe? Data scientists are working to make this more likely by matching personal data.

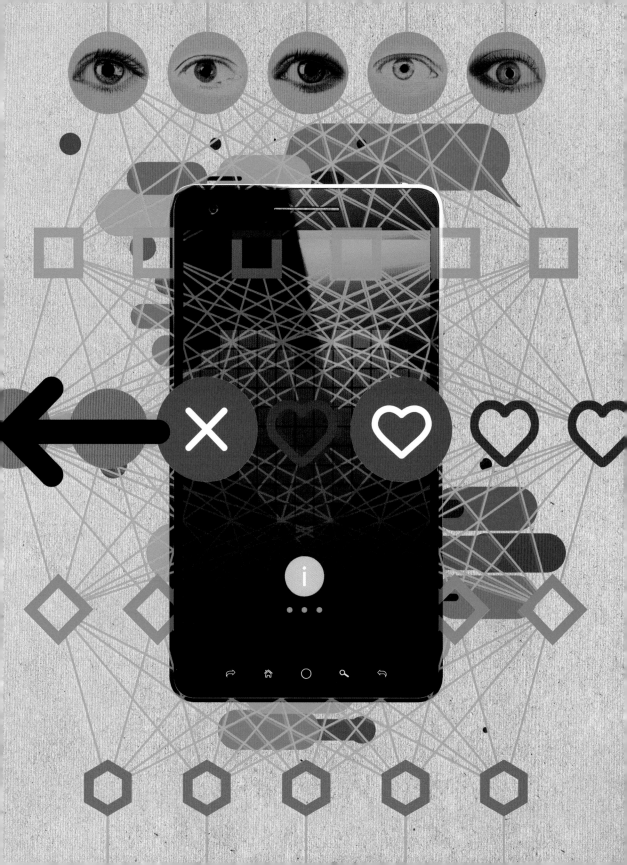

MUSIC
the 30-second data

The movement of music from physical libraries to digital libraries has changed the way we consume music. By having a digital music library, we have access to millions of songs by a variety of artists at the touch of a button. Given this volume of music, how are providers able to give us recommendations and custom playlists based upon our listening habits? Taking Spotify as an example, which is one of the most popular music streaming services in the world, it harnesses the power of data by adopting a three-pronged approach to determining what you might like. The first approach comes up with suggestions by comparing your listening habits to other users. The second approach applies machine-learning techniques to textual data such as news articles, blogs or even the text data stored within the digital music files themselves to find music you may like. The third approach analyses the raw audio content of the songs to classify similarity. Combining the results of these approaches allows music streaming services to come up with custom playlists for each and every user on the platform which can include a variety of genres and eras. Such streaming services are constantly evolving to harness new technologies.

RELATED TOPICS
See also
DATA COLLECTION
page 16

MACHINE LEARNING
page 32

3-SECOND BIOGRAPHIES
MARTIN LORENTZON &
DANIEL EK
1969– & 1983–
Swedish entrepreneurs who co-founded the popular music streaming service Spotify.

30-SECOND TEXT
Robert Mastrodomenico

3-SECOND SAMPLE
The digital age has opened up the world of music to us, but with so much choice, how can we find new music we like?

3-MINUTE ANALYSIS
If we consider two users: one may like songs a, b, c, d; the other a, b, c, e. Based on this, it could be suggested that the second user try song d and the first user song e, because they both like a, b and c. This is what is done on a much larger level for all music streaming service users.

Music streaming services can harness your listening data to introduce you to new music you might not have heard otherwise.

10 December 1815
Born in London

1816
Parents separate;
Lovelace is left in the
care of her mother

1824
Her father, Lord Byron,
passes away, aged 36

1829
Contracts the measles,
which leaves her
paralysed for a year

1833
Meets pioneering
mathematician Charles
Babbage for the first time

1835
Marries William King

1836
Becomes a mother for the
first time

1838
Becomes Countess of
Lovelace, following her
husband's earlship

1840s
Attempts to develop a
mathematical scheme for
gambling and nearly ends
up destitute

1842–3
Works on her translation
(with notes) on the
Analytical Engine

27 November 1952
Dies in London

ADA LOVELACE

Ada Lovelace is considered by many to be the founder of computer programming. A gifted mathematician – at a time when women were almost unheard of in STEM fields – she is said to have possessed a unique intuition into the workings of the analytical engine, the first digital computer. Her contributions to modern-day computer science are inescapable; her work either directly laid the foundations for or, at the very least, catalysed the ideas of many of the 'fathers' of modern computing, from Babbage to Turing.

Lovelace's mother, Lady Byron, ensured that her child grew up surrounded by mathematics tutors. Lady Byron was not a lover of mathematics, per se; rather, she saw maths and logic as a means to subdue her daughter's poetic disposition, a tendency that, she feared, Lovelace might have contracted from her father – none other than the poet Byron – despite never having had contact with him past infancy. Lovelace is said to have retained a fondness for her absent father, and her mathematical inventions were often imbued with a creativity that spoke of an artistic mind, not just a logical one.

Lovelace was introduced to the emerging world of computing machines at age 17, when she met Charles Babbage, who is generally credited as being the inventor of the first digital computer. She took a fascination to Babbage's work, and the two worked together over the next 20 years that comprised the remainder of Lovelace's short life. In the process, she helped craft the design and structure of the analytical engine, a complex and more powerful iteration of Babbage's difference engine that was supposed to, in Lovelace's words, 'weave... algebraic patterns, just as the Jacquard loom weaves flowers and leaves'.

While universally recognized as a central contributor to the analytical engine, Lovelace is better known for her translation of French engineer Luigi Menabrea's work on the analytical engine in 1843. Her translation was significantly longer than the original and contained many new ideas and improvements added in by Lovelace herself in the form of notes. This set of notes is supposed to have been a key inspiration for Alan Turing's first 'modern computer' nearly a century later.

Lovelace's imprint on modern computing is indelible. Even more remarkable is that she did so much in a life that spanned only 36 years, all the while balancing her roles as aristocrat, mother and mathematician.

Aditya Ranganathan

SPORTS

the 30-second data

Fields, courts and pitches have always welcomed professional and amateur statisticians alike to measure team and player performance. Common baseball metrics like Runs Batted In (RBI) and Earned Run Average (ERA) have been reliably recorded since the nineteenth century. Recent advancements in technology, however, have helped to launch a data science explosion felt by both participants and spectators. The invention of wearable technology has allowed data scientists to track athletes and activities. In tennis, for example, many professionals have turned to using racquets with embedded sensors, which allow them to track speed, rotation and ball hit location in real time. Other advancements include the expanded use of cameras and radar devices. In most major sports, the universal use of sophisticated cameras has provided participants and fans access to insights that were previously unimaginable. Using the new metrics provided by high-accuracy camera technology in Major League Baseball, for instance, team quantitative analysts have demonstrated that performance improves when batters correct their 'launch angle'. As athletes continue to use data science to improve their games, there is no sign that the ongoing arms race in sports metrics will get thrown out any time soon.

RELATED TOPICS
See also
LEARNING FROM DATA
page 20

STATISTICS & MODELLING
page 30

3-SECOND BIOGRAPHIES
BILL JAMES
1949–
Baseball writer and statistician who created a wide array of player evaluation formulas.

BILLY BEANE
1962–
Pioneered the use of non-traditional metrics for scouting undervalued baseball players.

30-SECOND TEXT
Scott Tranter

3-SECOND SAMPLE
While *Moneyball* is the most famous example, it pales in comparison to the advancements facilitated by the data revolution in sports, affecting the experiences of players, managers, referees and fans.

3-MINUTE ANALYSIS
Commentary surrounding the increase in data science's influence in sports often presents two competing factions: the purists and the nerds. The film *Moneyball* presented a heterodox, quantitatively-inclined baseball manager upending a system designed by experienced scouts. Recently, some athletes have criticized statisticians for their lack of experience. However, the most successful data teams leverage expert insights to complement their analytics.

Sports data science works best when integrating athlete experience and scientists' numbers.

SOCIAL MEDIA

the 30-second data

3-SECOND SAMPLE
Since the beginning of the twenty-first century, social media has taken over how humans interact, access news and discover new trends, driven, in large part, by data science.

3-MINUTE ANALYSIS
TV shows such as *Black Mirror* provide us with a different view towards continuing advancements in social media. The episode 'Nosedive' depicts a world where 'social credit', deriving from a mixture of in-person and online interactions, dictates where a person can live, what they can buy, who they can talk to and more. China has begun implementing a Social Credit System to determine the trustworthiness of individuals and accept or deny individuals for functions such as receiving loans and travelling.

In just a few years, companies like Facebook, Snapchat and Twitter went from small internet start-ups to multibillion-pound tech giants with priceless quantities of influence. Facebook is now reaching nearly 90 per cent of all mobile users in the US, while Twitter boasts 100 million daily active users, enough to become the fifteenth most populous country in the world. With such a massive user outreach, companies can leverage the enormous amounts of data generated on these platforms to discover new trends and insights about their audience, then apply this knowledge to make smarter, more reliable business decisions. By tracking users and implementing algorithms to learn their interests, social media companies have been able to deliver highly targeted adverts and generate billions of pounds in ad revenue every year. These same machine-learning algorithms can be used to tailor the content each user sees on their screen. From a timeline to suggested friends, social media companies play a prominent role in how users interact with their apps and, subsequently, the world around them. What once started as a way to update friends on one's status has evolved into a public forum, marketplace and news outlet all rolled into one.

RELATED TOPICS
See also
PRIVACY
page 86

MARKETING
page 108

TRUSTWORTHINESS SCORE
page 146

3-SECOND BIOGRAPHIES
JACK DORSEY
1976–
Co-founder and CEO of Twitter.

MARK ZUCKERBERG
1984–
Co-founder and CEO of Facebook, and the youngest self-made billionaire, at 23.

30-SECOND TEXT
Scott Tranter

The rapid growth of social media has seen it infiltrate everyday life, with data capture capabilities on an unprecedented scale.

GAMING

the 30-second data

Competitive video gaming, known as esports, is an emerging global phenomenon in which professional players compete in packed stadiums for prize pools reaching millions of pounds. Unlike with traditional sporting events, esports fans engage more directly with content via online live streaming technology on platforms such as Twitch. Esports consumers largely consist of males in the 20 to 30 age range, a prime demographic that companies wish to target. By tracking the fan base's habits and interests using analytical tools and survey methods, companies have been able to tailor content based on the audience they wish to target. However, because of the esports audience's reduced television consumption and tendency to block internet adverts using browser-based ad-blocking technology, companies are looking into non-traditional methods to reach this demographic. For example, due to the digital nature of esports, brands have the ability to display their products directly in the video games, avoiding ad-blockers altogether. Additionally, professional esports players have a large influence on how their fans may view certain products. To take advantage of this, companies often partner with these influencers and utilize their popularity in order to reach target audiences for their products.

RELATED TOPICS
See also
LEARNING FROM DATA
page 20

SPORTS
page 126

3-SECOND SAMPLE
Esports is engaging its young, digital-savvy fans through non-traditional online media, paving the way for data science into the recreational trends of young generations.

3-MINUTE ANALYSIS
Although esports thrived off the back of online live streaming technology, it has also begun broadcasting esports on television, with ads displaying during commercial breaks, akin to traditional sports. Esports companies are adopting the geolocated franchising model, which looks to take advantage of television advertising and sponsorship deals for its revenue. With this move, esports has an opportunity to expand its reach, opening up the door for mainstream popularity.

3-SECOND BIOGRAPHIES
JUSTIN KAN
1983–
American internet entrepreneur who co-founded Twitch, formerly Justin.tv, the most popular streaming platform for esports content.

TYLER 'NINJA' BLEVINS
1991–
American Twitch streamer, internet personality and former professional gamer who helped bring mainstream attention to the world of esports.

30-SECOND TEXT
Scott Tranter

As the esports industry grows, top players may soon be able to sign endorsement deals in the ballpark of professional athletes.

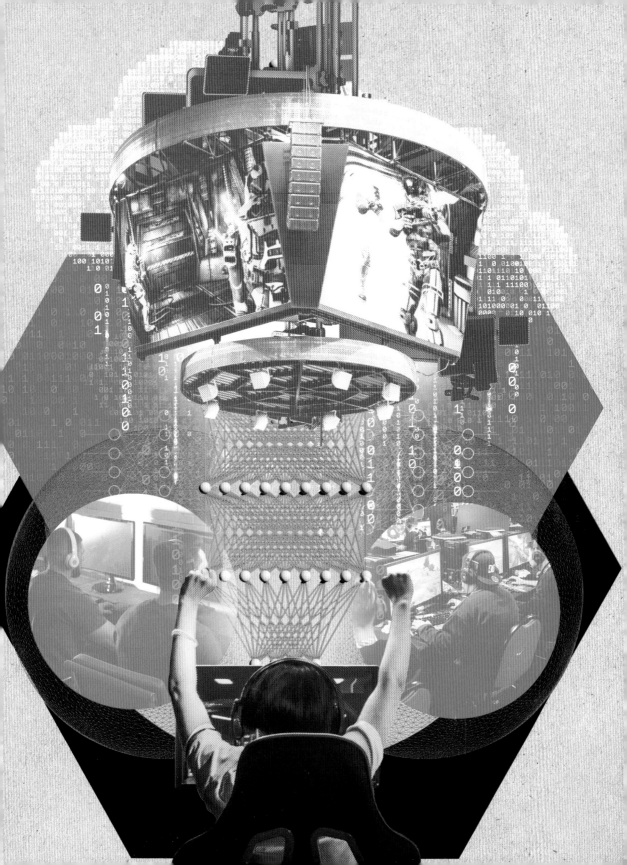

GAMBLING

the 30-second data

In gambling, everything from the likelihood that the dealer will bust in blackjack to the placement of specific slot machines at key locations are driven by statistics. And, in the evolving world of data science, those with greater access to it can find themselves at a huge advantage over others. This ranges from the simple approach of an experienced poker player understanding the odds of turning his straight-draw into a winning hand – and the correlated risk of pursuing that potential hand – to the more advanced techniques casinos use to turn vast amounts of unstructured data into predictions on the best way to entice players to bet, and to bet more, on lower-odd payouts. Resources exist for both the house and for the player, and they extend well beyond card games and slot machines. Statistical models can impact the payout of sports events – oftentimes adjusting odds in real time and based on the direction that money is moving – in a way that can minimize the risk of the sportsbook (the part of casinos that manages sports betting). By the same token, some gamblers use or create statistical models to make educated decisions on outcomes that are data-driven rather than narrative-driven, giving them an edge on those following their instinct.

RELATED TOPICS

See also
LEARNING FROM DATA
page 20

SURVEILLANCE
page 82

SPORTS
page 126

3-SECOND BIOGRAPHIES
RICHARD EPSTEIN
1927–
Game theorist who has served as an influential statistical consultant for casinos.

EDWARD O. THORP
1932–
Mathematician who pioneered successful models used on Wall Street and in casinos.

30-SECOND TEXT
Scott Tranter

3-SECOND SAMPLE
Data science and gambling can blend together with devastating effect – and has made the adage 'the house always wins' even more true.

3-MINUTE ANALYSIS
There have been reports on the ways in which casinos are utilizing decades' worth of player data (tied back to individual players through their rewards cards), while plenty of 'expert' gamblers have written books designed to 'beat the house'. Those with designs on gambling based on luck are simply playing the wrong game – they should be playing the stats – while hoping that Lady Luck still shines on them.

Move over Lady Luck: professional gamblers now pit their data skills against those of the house.

THE FUTURE ◑

AI (artificial intelligence) Often used interchangeably with 'machine learning'. The process of programming a computer to find patterns or anomalies in large data sets, or to find the mathematical relationship between some input variables and an output.

box plot Data visualization which shows the distribution, or shape, of a data set. It includes important descriptive statistics including the median, lower and upper quartiles, and any outliers. This is sometimes referred to as a 'box and whisker plot'.

chat bot Computer program designed to interact or 'chat' with human users through text messaging; often used in customer service, to deal with customers' questions and problems more efficiently than a human could.

data ethics Concerned with how data can be ethically collected, stored, analysed and distributed; especially important when handling personal data.

data legality Legislation concerned with how data can be collected, stored and accessed, and by whom.

deepfake An image, video or audio clip that has been altered using AI to show a person doing or saying something they have not done. This may include superimposing a person's head on the body of another person in an image, or a speech track over a video of a person's mouth moving.

genomics The study of the structure and function of DNA.

longitudinal behavioural surveys Study that takes repeated measurements from the same subjects or participants over time.

machine learning Finding a mathematical relationship between input variables and an output. This 'learned' relationship can then be used to output predictions, forecasts or classifications given an input.

mine information Collecting data, typically from the internet, on a large scale. Data can be mined or 'scraped' directly from websites. There are important ethical and privacy considerations when mining personal data.

nanotechnologies Technologies that exist at nanometre scale, and often involve the manipulation of individual molecules. This includes nanomaterials such as carbon nanotubes, and 'buckyballs' or Buckminsterfullerene.

self-learning Type of machine learning, commonly used to find patterns or structure in data sets. Also known as 'unsupervised learning'.

smart Refers to internet-connected devices with real-time analysis or machine learning capabilities. Smart watches typically include physical activity monitors and internet connectivity, and smart TVs may include voice recognition.

social scoring System whereby citizens are 'scored' by government agencies according to their behaviour, which may include adherence to the law, financial stability, employment status or educational attainment. A person's social score may affect their ability to access social services or loans.

stem-and-leaf display A form of data visualization, similar to a histogram, used to show the distribution or shape of a data set.

time series analysis The analysis of a signal or variable that changes over time. This can include identifying seasonal trends or patterns in the data, or forecasting future values of the variable.

topology Branch of mathematics concerned with geometric objects and their properties when they are stretched, twisted or crumpled.

PERSONALIZED MEDICINE

the 30-second data

Humans have always been

interested in themselves. So it's no surprise that they want to know what's happening in their bodies – at all times. Consumer demand for personalized health data has fuelled the success of smart watches, fitness trackers and other wearable devices which give real-time feedback. But what does the next generation of wearables look like? And what can the data tell us? With technology so deeply ingrained in our lives, it is easy to imagine a future with technologically advanced clothing, smart skin patches or ingestible nanotechnologies which detect or monitor disease. Instead of a one-off blood test, we could all be wearing a smart patch made of a series of micro-needle sensors that continually track chemical changes under the skin. Or flexible and stretchable sensors resembling tattoos, which could monitor lactate during a workout or sense changes in environmental chemicals and pollutants. And imagine the data – huge amounts of data. Future wearable technology will collect thousands of data points a minute, maybe even a second, which will need powerful algorithms, machine learning and AI to reduce the data into meaning. This will be essential to mine the information, to better understand disease, population-wide health trends and the vital signs to predict a medical emergency.

3-SECOND SAMPLE
Wearable technology could tap in to huge amounts of human data, opening up the possibility of real-time healthcare, along with new ways to detect and prevent disease.

3-MINUTE ANALYSIS
As the line between consumer wearables and medical devices blurs, concerns are rising about the reliability and security of data. For example, current smartphone apps for melanoma skin cancer detection have a high failure rate. If people are deciding to change their lifestyle, or medical professionals are making treatment decisions, it is crucial that wearable devices go through any necessary clinical trials and are supported by strong scientific evidence.

RELATED TOPICS
See also
EPIDEMIOLOGY
page 76

ETHICS
page 152

3-SECOND BIOGRAPHIES
JOSEPH WANG
1948–
American researcher and director of the Center for Wearable Sensors at University of California, San Diego, who is pioneering wearable sensors to monitor disease.

JEFF WILLIAMS
1963–
Apple's chief operating officer, who oversees the Apple watch and the company's health initiatives.

30-SECOND TEXT
Stephanie McClellan

To attain secure and reliable data, the personal healthcare industry needs to be properly regulated.

MENTAL HEALTH

the 30-second data

Mental health disorders affect
over 970 million people worldwide, tend to be
under-diagnosed, can have long-term effects
and often carry social stigma. Mental health
data involves longitudinal behavioural surveys,
brain scans, administrative healthcare data and
genomics research. Such data is difficult to
obtain and of a sensitive nature. Data science
facilitates access to this data and its applications
to mental health include virtual counselling,
tele-psychiatry, effective use of social media
and analysing mobile health device data. Data
science interfaces with mobile applications
to track patient mood with visual devices such
as flowcharts, produce algorithms to generate
targeted advice and connect people to a
therapist in order to improve symptoms of
depression and anxiety, for example. Machine
learning mines unstructured text from social
media and similar sources to detect symptoms
of mental illness, offers diagnoses and creates
algorithms to predict the risk of suicide.
Furthermore, AI can combine brain scans
and neural networks to deliver personalized
psychiatric medicine. However, preserving
confidentiality of personal information,
maintaining data security and transparency
with the use of data are key.

3-SECOND SAMPLE
Data science enables
digital mental healthcare,
to improve access and
treatment outcomes.

3-MINUTE ANALYSIS
Mental healthcare has been
made more accessible
to overcome provider
shortages and stigma via
tele-psychiatry and virtual
agents. A 3D chat bot
named Ellie has the
capability to show facial
expressions and detect
non-verbal cues. Patients'
verbal input and facial
expressions run through
algorithms, which
determine Ellie's visual
and verbal responses.
In a study, Ellie was more
effective in diagnosing
post-traumatic stress
disorder among military
personnel during routine
health exams. The ability
to build rapport and trust
with patients drives Ellie's
effectiveness.

RELATED TOPICS
See also
NEURAL NETWORKS &
DEEP LEARNING
page 34

HEALTH
page 92

PERSONALIZED MEDICINE
page 138

3-SECOND BIOGRAPHIES
EMIL KRAEPELIN
1856–1926
Used data science concepts
in his diagnosis methods for
schizophrenia and bipolar.

AARON BECK
1921–
Created the Beck Depression
Inventory, used to diagnose
depression.

30-SECOND TEXT
Rupa R. Patel

*The application
of data science to
mental health carries
important ethical
considerations.*

SMART HOMES

the 30-second data

Our homes are growing smarter, from turning the lights on, to playing a song or even ordering a pizza, devices that have become part of the furniture of the modern home can now do it all. Actions that were fully manual are controlled via a variety of methods, be it by phone or smart speaker, so the central heating can be turned on at the touch of a button or the sound of a voice. The more smart devices are used in the home, the more data they emit and the more information that becomes available to collect. The volume of data lends itself well to applying machine learning and AI techniques. In the case of smart speakers, for example, once you've asked for something, it is sent over the internet to be processed by a voice-processing service, which does all the hard work. The more data the speaker receives, the better it becomes at processing a request. This means that a smart home should grow increasingly smarter. One worry is how they learn: major companies have had to apologize for employing people to listen in to conversations in private homes with the alleged aim to improve voice recognition – valid concerns over privacy loom.

3-SECOND SAMPLE
Machine learning and AI have now taken the average home and turned it into a smart home.

3-MINUTE ANALYSIS
Smart speakers now use self-learning technology, which means that they can learn from the errors they make. So if you ask for something that the speaker doesn't understand and then ask again in a different way, it can learn that the first request is related to the second, combine the two and learn from them both.

RELATED TOPICS
See also
TOOLS
page 22

MACHINE LEARNING
page 32

ARTIFICIAL INTELLIGENCE (AI)
page 148

3-SECOND BIOGRAPHIES
WILLIAM C. DERSCH
fl. 1960s
Creator of IBM's Shoebox, the first known voice assistant.

ROHIT PRASAD & TONI REID
fl. 2013–
Creators of the Alexa assistant and AI specialists at Amazon.

30-SECOND TEXT
Robert Mastrodomenico

'Smart' stands for Self-Monitoring Analysis and Reporting Technology, and such devices link to the Internet of things.

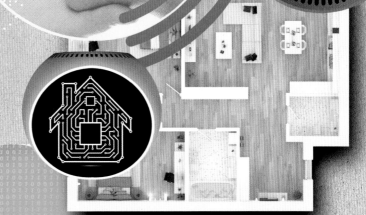

16 June 1915
Born in Bedford,
Massachusetts, USA

1933
Enters school for the first
time, attending Brown
University

1936
Receives Bachelor's
degree in Chemistry

1939
Receives PhD in
Mathematics at Princeton
University, and is
appointed to the
mathematics faculty

1945
Begins a 40-year
association with AT&T
Bell Laboratories

1960
Begins a two-decade
association with NBC,
working on election
night forecasting and
developing new methods
for avoiding erroneous
judgements. For several
years Tukey can be seen
walking around in the
background of the
NBC set

1965
Becomes the founding
chairman of Princeton's
new Department of
Statistics

1973
Awarded the US National
Medal of Science; advises
every US President from
Eisenhower to Ford

1985
Retires from Princeton
and Bell Laboratories,
and delivers valedictory
address with the title
'Sunset Salvo'. Continues
to advise many boards
and governmental
councils

26 July 2000
Dies in New Brunswick,
New Jersey, USA

JOHN W. TUKEY

John W. Tukey was born in 1915, and he showed unusual brilliance from a very early age. He was schooled at home by his parents until he entered Brown University, Rhode Island. He graduated in three years with BA and MS degrees in Chemistry. From Brown he moved on to Princeton, where he started in Chemistry but soon switched to Mathematics, getting a PhD at age 24 in 1939 with a dissertation in Topology, and he then moved directly to a faculty position at the same university. He remained at Princeton until he retired in 1985, adding a part-time appointment at Bell Telephone Laboratories in 1945.

Over a long career, Tukey left his imprint in many fields. In mathematics: Tukey's formulation of the axiom of choice; in time series analysis: the Cooley–Tukey Fast Fourier Transform; in statistics: exploratory data analysis, the jackknife, the one-degree test for non-additivity, projection pursuit (an early form of machine learning), and Tukey's method of guaranteeing the accuracy of a set of simultaneous experimental comparisons. In data analysis alone he created an array of graphical displays that have since become standard, widely used to facilitate the discovery of unsuspected patterns in data and to learn about distributions of data, including the box plot and the stem-and-leaf display. He coined other new terms that also became standard terminology, including 'bit' (short for 'binary digit') for the smallest unit of information transferred, and even more terms that did not catch on (virtually no one remembers his 1947 term for a reciprocal second: a 'whiz').

It was through his teaching and consistent emphasis on the importance of exploratory data analysis as a basic component of scientific investigation that Tukey was established as a founder of what some today call data science, and some credit him with coining the term. From 1960 to 1980, he worked with the television network NBC as part of an election night forecasting team that used early partial results to 'call' the contested races of interest, working with several of his students, including at different times David Wallace and David Brillinger. In 1960 he prevented NBC from prematurely announcing the election of Richard Nixon. Tukey's favourite pastimes were square dancing, bird watching and reading several hundreds of science fiction novels. He died in 2000 in New Brunswick, New Jersey.

Stephen Stigler

TRUSTWORTHINESS SCORE

the 30-second data

The idea of being judged in
society based on your behaviour is not new, but
the data age is giving it all new possibilities and
meaning. China is the first to take advantage of
this on a population scale, by giving each of its
citizens a social credit score, which can move
up or down, based upon AI and data collection
techniques of each individual's behaviour.
The methodology of the score is a secret,
but examples of infractions are bad driving
(captured by China's extensive camera systems),
buying too many video games and posting
deemed 'fake news' online. The result? A low
score will be punishable by actions such as
travel bans, not allowing your child into certain
schools, stopping you from getting the best
jobs or staying at the best hotels, slowing of
your internet and, at its most base level, a loss
of social status (including publishing your score
with your online dating profile). Other countries
and companies are on the brink of using systems
very similar to this. The UK, for example, has
been accused of getting quite close with the
usage of data from sources such as credit
scores, phone usage and even rent payments
to filter access to many social services and jobs.

RELATED TOPICS
See also
SURVEILLANCE
page 82

PRIVACY
page 86

ETHICS
page 152

3-SECOND BIOGRAPHY
EARL JUDSON ISAAC
1921–83
Mathematician who, along
with Bill Fair, founded a
standardized, impartial
credit scoring system.

30-SECOND TEXT
Liberty Vittert

3-SECOND SAMPLE
Yell at someone in traffic
who cuts in front of you?
Your social credit score
might go down: a data-
driven score based on
individual behaviour (a
black box of lifestyle traits).

3-MINUTE ANALYSIS
Companies are not immune
to interest in a social score.
Facebook reportedly
assigns all of its users a
trustworthiness score,
with the intention to
combat the spread of
misinformation. In this
case, the score itself and
how it is computed are a
complete secret. In 2016,
the online profiles of over
70,000 OKCupid users
(including usernames,
political interest, drug
usage and sexual exploits)
were scraped and published
by academic researchers.

*The amount of public
and private data on
individuals is endless,
so where will our social
scoring end?*

ARTIFICIAL INTELLIGENCE (AI)

the 30-second data

3-MINUTE ANALYSIS
The full impact of AI on the labour market is unknown, and advances such as driverless cars may lead to unemployment or a need to re-skill. However, many automated systems will always require a degree of human oversight or interpretation, particularly in sensitive fields such as medicine or criminal justice.

Artificial Intelligence is less about sentient robots and more about computers that can find patterns and make predictions using data. You encounter AI every time your phone autocompletes your messages, whenever you use a voice recognition service or when your bank catches a fraudulent transaction on your credit card. On many social media platforms, facial recognition technology uses AI to establish whether an uploaded image contains a face, and to identify the subject by matching against a database of images. At airports, this same technology is increasingly used to identify travellers and perform passport checks. However, facial recognition algorithms are notorious for performing poorly on darker skinned faces, so a technology that saves time at passport control for one person could misidentify another traveller as a wanted criminal. As well as performing image recognition tasks, AI can be used to generate hyper-realistic images of faces or landscapes. This has benefitted the computer graphics and video game industries, but has also enabled the rise of the 'deepfake', an image or video in which a person is shown doing or saying something they haven't done. As deepfakes become ever more convincing, governments and news outlets must find ways to combat this new form of misinformation.

RELATED TOPICS
See also
MACHINE LEARNING
page 32

BIAS IN ALGORITHMS
page 50

ETHICS
page 152

3-SECOND BIOGRAPHY
ALAN TURING
1912–54
British mathematician and computer scientist, widely considered the father of AI.

30-SECOND TEXT
Maryam Ahmed

Facial recognition is one of the most widespread uses of AI in everyday life.

REGULATION

the 30-second data

Advances in data science are raising new questions that have politicians and legislators scratching their heads. How can personal data be collected and stored ethically and securely? Who should check whether algorithms are biased? And whose fault is it when the irresponsible use of data causes real-life harm? In Europe, the General Data Protection Regulation (GDPR) gives citizens control over their data; organizations must ask for permission to collect personal data, in simple language rather than legal-speak, and delete it when asked. The European Commission recommends that AI should be transparent and unbiased with human supervision if needed, and in the US the Algorithmic Accountability Act will soon require that AI systems are ethical and non-discriminatory. In the wake of high-profile data privacy scandals at Facebook and DeepMind, some companies have made efforts to self-regulate, with varying success. Microsoft and IBM have publicly committed to building fair and accountable algorithms, with data privacy and security as key concerns. Meanwhile, Google has explicitly pledged not to develop algorithms for weapons systems, or technologies that contravene human rights, although its AI ethics board was dissolved just one week after being founded, due to board member controversies.

3-SECOND SAMPLE
Technology is evolving too fast for lawmakers to keep up with, so who should be responsible for regulating the use of data science and AI?

3-MINUTE ANALYSIS
NGOs can hold algorithms to account and effect change where governments fail. The Algorithmic Justice League in the US has exposed racial and gender biases in IBM and Microsoft's face recognition systems, and in the UK, Big Brother Watch has found low accuracy rates in face recognition algorithms used by some police forces. Algorithm Watch has investigated an opaque credit scoring algorithm in Germany, demonstrating that some demographic groups are unfairly penalized.

RELATED TOPICS
See also
BIAS IN ALGORITHMS
page 50

IBM'S WATSON &
GOOGLE'S DEEPMIND
page 94

ARTIFICIAL INTELLIGENCE (AI)
page 148

3-SECOND BIOGRAPHY
CORY BOOKER
1969–
US Senator (Democrat) and sponsor of the Algorithmic Accountability Act.

30-SECOND TEXT
Maryam Ahmed

The main challenge for data regulators is to keep up with the fast pace of technological developments.

ETHICS

the 30-second data

Data is the new oil, except it isn't the earth that is being excavated – it's you. Your data is being used as a commodity. It is being bought, sold and potentially stolen, from the Weather Channel selling your geo-located data, to political advertising companies matching voter registration records to Facebook profiles, companies sorting through CVs based upon algorithmic functions, or insurance companies selling your private medical information to third parties to contact you about 'further treatment'. With this new industry comes overwhelming ethical concerns. Both the private and public sector, behind closed doors, must make very difficult ethical decisions about what they can and cannot do with your data. These questions arise not only in the pursuit of making money, but also in an effort to mitigate and solve serious problems. For example, should Apple, or other electronic/technology companies, give the government access to the data on the mobile phone of an alleged terrorist? Data ethics encompasses the collection, analysis and dissemination of data (including who owns it), as well as any associated algorithmic processes and AI. Professional statistical bodies and companies alike have tried to convene thought groups on the many quandaries surrounding data ethics.

RELATED TOPICS
See also
PRIVACY
page 86

REGULATION
page 150

3-SECOND BIOGRAPHIES
RUFUS POLLOCK
1980–
Founder of the Open Knowledge Foundation, which promotes and shares open data and content.

EDWARD SNOWDEN
1983–
American whistle-blower of classified National Security Agency information.

30-SECOND TEXT
Liberty Vittert

3-SECOND SAMPLE
Data ethics seems antithetical: $2+2=4$, what can be unethical about that? But the saying, 'figures never lie, but liars always figure' isn't really that true.

3-MINUTE ANALYSIS
From the leaks by Edward Snowden to the extent that companies have an ethical or legal responsibility to inform users of how their data will be used, the conflation of data ethics versus data legality is a complicated issue. In the Cambridge Analytica scandal, while their business practice was questionable, thousands, if not tens of thousands, of other app developers were using the features that Facebook itself created, to do precisely the same thing that Cambridge Analytica was doing. What is ethical and which is the responsible party?

Data ethics is an ever-changing discipline based upon current practices.

RESOURCES

BOOKS & ESSAYS

A First Course in Machine Learning
S. Rogers & M. Girolami
Chapman and Hall/CRC (2016)

*An Accidental Statistician: The Life
and Memories of George E.P. Box*
G.E.P. Box
Wiley (2013)

Alan Turing: The Enigma
Andrew Hodges
Vintage (2014)

The Art of Statistics: How to Learn from Data
David Spiegelhalter
Pelican Books (2019)

*The Book of Why: The New Science
of Cause and Effect*
Judea Pearl & Dana Mackenzie
Allen Lane (2018)

'Darwin, Galton and the Statistical
Enlightenment'
S. Stigler
Jour. of the Royal Statist. Soc. (2010)

Data Science for Healthcare
Sergio Consoli, Diego Reforgiato
Recupero, Milan Petković (Eds)
Springer (2019)

The Elements of Statistical Learning
J. Friedman, T. Hastie & R. Tibshirani
Springer (2009)

Get Out the Vote!
Donald P. Green & Alan S. Gerber
EDS Publications Ltd (2008)

Healthcare Data Analytics
Chandan K. Reddy, Charu C. Aggarwal (Eds)
Chapman & Hall/CRC (2015)

Invisible Women
Caroline Criado Perez
Chatto & Windus (2019)

'John Wilder Tukey 16 June 1915–26
July 2000'
P. McCullagh
*Biographical Memoirs of Fellows of the Royal
Society* (2003)

*Machine Learning: A Probabilistic
Perspective*
K.P. Murphy
MIT Press (2012)

The Mathematics of Love
Hannah Fry
Simon & Schuster (2015)

Memories of My Life
F. Galton
Methuen & Co. (1908)

*Naked Statistics: Stripping the Dread
from the Data*
Charles Wheelan
W.W. Norton & Company (2014)

The Numerati
Stephen Baker
Mariner Books (2009)

Pattern Recognition and Machine Learning
C.M. Bishop
Springer (2006)

*The Practice of Data Analysis: Essays in
Honour of John W. Tukey*
D. Brillinger (Ed)
Princeton Univ. Press (1997)

*Statistics Done Wrong: The Woefully
Complete Guide*
Alex Reinhart
No Starch Press (2015)

The Victory Lab
Sasha Issenberg
Broadway Books (2013)

WEBSITES

Coursera
www.coursera.org/learn/machine-learning

Data Camp
www.datacamp.com/courses/introduction-
to-data

The Gender Shades project
gendershades.org
Uncovered bias in facial recognition
algorithms

ProPublica
www.propublica.org/article/machine-bias-
risk-assessments-in-criminal-sentencing
Investigated the COMPAS algorithm for
risk-scoring prisoners

Simply Statistics
simplystatistics.org

Udemy
www.udemy.com/topic/data-science/

NOTES ON CONTRIBUTORS

EDITOR

Liberty Vittert is a Professor of the Practice of Data Science at the Olin Business School at the Washington University in St Louis. She is a regular contributor to many news organizations as well as having a weekly column "A Statistician's Guide to Life" on Fox Business. As a Royal Statistical Society Ambassador, BBC Expert Woman and an Elected Member of the International Statistical Institute, Liberty works to communicate statistics and data to the public. She is also an Associate Editor for the *Harvard Data Science Review* and is on the board of USA for UN Refugee Agency (UNHCR) as well as the HIVE, a UN Refugee Agency data initiative for refugees.

FOREWORD

Xiao-Li Meng is the Whipple V. N. Jones Professor of Statistics at Harvard University, and the Founding Editor-in-Chief of *Harvard Data Science Review*. He was named the best statistician under the age of 40 by COPSS (Committee of Presidents of Statistical Societies) in 2001, and served as the Chair of the Department of Statistics (2004–12) and the Dean of Graduate School of Arts and Sciences (2012–17).

CONTRIBUTORS

Maryam Ahmed is a data scientist and journalist at BBC News, with a PhD in Engineering from the University of Oxford. She has reported on issues such as targeted political advertising and the gender pay gap. Maryam is a strong advocate of transparency in the public sphere, and has spoken on this topic at venues including the Royal Society of Arts.

Vinny Davies completed his PhD in Statistics before becoming an academic researcher in Machine Learning. He has spent most of his career looking at applications of probabilistic models in Biology and Chemistry, including models for vaccine selection and the left ventricle of the heart.

Sivan Gamliel, Director, is a member of BlackRock Alternative Advisors, the firm's hedge fund solutions team, where she serves as Head of Quantitative Strategies. She received her Bachelor of Science degree in Physics from Massachusetts Institute of Technology, USA.

Rafael A. Irizarry is professor and chair of the Department of Data Sciences at Dana-Farber Cancer Institute and Professor of Biostatistics at Harvard T.H. Chan School of Public Health.

Robert Mastrodomenico PhD (University of Reading) is a data scientist and statistician. His research interests revolve around the modelling of sport events and computational techniques, with specific focus on the Python programming language.

Stephanie McClellan is a science writer based in London, and has an MSc in Science Communications from Imperial College London. She has been a freelance writer for institutions such as the European Space Agency, the BBC, CERN, and the United Nations Educational, Scientific and Cultural Organization (UNESCO). She has also worked in the national press office at Cancer Research for five years.

Regina Nuzzo has a PhD in Statistics from Stanford University and graduate training in Science Writing from University of California Santa Cruz. Her writings on probability, data and statistics have appeared in the *Los Angeles Times*, *New York Times*, *Nature*, *Science News*, *Scientific American* and *New Scientist*, among others.

Rupa Patel is a physician scientist and is the Founder and Director of the Washington University in St Louis Biomedical HIV Prevention programme. She is also a technical advisor for the World Health Organization. Dr Patel utilizes data science to improve implementation of evidence-based HIV prevention strategies in clinics, health departments and community organizations in the US, Africa and Asia.

Aditya Ranganathan is the chief evangelist for Sense & Sensibility and Science (S&S&S), a UC Berkeley Big Ideas course – founded by Saul Perlmutter – on critical thinking, group decision making and applied rationality. He also serves on the board of Public Editor, a citizen science approach to low-quality news and fake news. Aditya is pursuing his PhD at Harvard University, where he studies collective behaviour (with implications for group dynamics and education).

Willy C. Shih is the Robert & Jane Cizik Professor of Management Practice at the Harvard Business School. He worked in the computer and consumer electronics industries for 28 years, and has been at the school for 13 years.

Stephen M. Stigler is the Ernest DeWitt Burton Distinguished Service Professor of Statistics at the University of Chicago. Among his many published works is 'Stigler's Law of Eponymy' ('No scientific discovery is named after its original discoverer' in *Trans. N. Y. Acad. Sci.* 1980, 39: 147–158). His most recent book on the history of statistics is *The Seven Pillars of Statistical Wisdom* (2016).

Scott Tranter is the former Director of Data Science for Marco Rubio for President and founder of Øptimus, a data and technology company based in Washington, DC. Tranter has worked in both the political and commercial spaces where the science of using data to innovate how we do everything from elect our leaders to sell people cars has been evolving over the last several decades.

Katrina Westerhof helps companies develop and adopt emerging technologies, particularly in spaces that are being upended by analytics and the Internet of things. She has a diverse background in consulting, innovation, engineering and entrepreneurship across the energy, manufacturing and materials industries.

INDEX

ACKNOWLEDGEMENTS

The publisher would like to thank the following for permission to reproduce copyright material on the following pages:

All images that appear in the montages are from **Shutterstock, Inc**. unless stated.

Alamy Stock Photo/Photo Researchers: 106

Getty Images/Donaldson Collection: 124; Alfred Eisenstaedt: 144

Library of Congress: 51, 91

NASA/CXC/RIKEN/T. Sato et al: 65

North Carolina State University/College of Agriculture and Life Sciences, Department of Communication Services Records (UA100.099), Special Collections Research Center at North Carolina State University Libraries: 70

Wellcome Collection/John Snow: 21; Wellcome Images: 26, 88

Wikimedia Commons/ARAKI Satoru: 47; Cdang: 25; CERN: 63; Chabacano: 57; Chrislb:6, 35, 85; David McEddy: 52; Denis Rizzoli: 91; Emoscopes: 49; Fanny Schertzer: 127; Fred053: 127; Geek3: 65; Headbomb: 127; Justinc: 21; Karsten Adam: 127; Martin Grandjean: 2,29; Martin Thoma: 6, 35; MLWatts: 49; National Weather Service: 103; Niyumard: 133; Paul Cuffe: 103; Petr Kadlec: 127; Sigbert: 25; Trevor J. Pemberton, Michael DeGiorgio and Noah A. Rosenberg: 69; Tubas: 127; Warren K. Leffler: 51; Yapparina: 133; Yomomo: 43; Yunyoungmok: 25; Zufzzi: 33

All reasonable efforts have been made to trace copyright holders and to obtain their permission for the use of copyright material. The publisher apologizes for any errors or omissions in the list above and will gratefully incorporate any corrections in future reprints if notified.